Your Passport to a Career in Bioinformatics

Prashanth N. Suravajhala

Your Passport to a Career in Bioinformatics

 Springer

Prashanth N. Suravajhala
Bioclues.org
Secunderabad
Andhra Pradesh
India

ISBN 978-81-322-1162-4   ISBN 978-81-322-1163-1   (eBook)
DOI 10.1007/978-81-322-1163-1
Springer New Delhi Heidelberg New York Dordrecht London

Library of Congress Control Number: 2013939043

Printed on acid-free paper

Springer is part of Springer Science+Business Media (www.springer.com)

*To my Mother
Nirmala Sastry*

Foreword

When I first heard about the field of bioinformatics, I was a university senior majoring in chemistry. It was 1995, and my intention at the time was to focus on the application of chemistry in the life sciences. In fact, in those days I was interested in any field of science or engineering that could be applied to biology.

But, when it came time to select a project for my senior thesis, I was asked by my thesis adviser if I had an interest in computers. Certainly, I did. I had a year of computer science courses under my belt, but I also had an avid interest in computers as a hobby—I wrote my first BASIC program circa 1981 on a friend's Atari 800. And, so my adviser proceeded to tell me that there is this nascent field called "bioinformatics", which is a hybrid of computer science and biology.

I immediately fell in love with the idea that I could combine a professional interest of mine with a personal one. And, from then on, even through graduate school, all of my research projects involved programming. Not one required that I stand at a bench with a micropipette, as I knew I would be doing as a biochemist.

Of course, it did not go over so well with many of the professors back then that a student would pursue a degree in either biochemistry or biology with a purely computational project. In the 1990s, there were just a handful of degree programs in bioinformatics in the whole world—one of them halfway around the world from where I lived. But, I limited my own geographical options, and it seemed that my only choice was to pursue a graduate degree in "traditional" biochemistry and find an adviser and laboratory group that had an interest in performing computational analyses on their data.

Fortunately for aspiring scientists today, there are many straightforward ways to enter the field of bioinformatics. To that point, there are scores of degree programs throughout the world—many of them online degrees.

And, there are other ways to further one's own career as a bioinformatics practitioner. For one, there is the Bioinformatics.Org website, of which I am the founder, with Prashanth Suravajhala among the directors. Prash also founded Bioclues.org and has been active in mentoring students online regarding their academic projects in bioinformatics. It is because of this experience of his, that I think you will be enlightened by the insight that Prash shares within these pages.

J. W. Bizzaro
President, Bioinformatics.org

Prologue

Today we define success by publicity and bank accounts. But that's not really success at all. Don't believe the hype. Success is ephemeral. You have to define it yourself.

Chris North

Most people would succeed in small things if they were not troubled with great ambitions.

Henry Wadsworth Longfellow

Any new word invites inquiry, excitement and sometimes disdain and so was bioinformatics, at least in developing countries. Theoretical bioinformatics although born in the 1980s, has flourished ever since, as many new academic and empirical developments with focal point on wet-lab research confirm. Bioinformatics is now regarded as a tool but fantasized as a familiar science even by few scientists who have had track record of early career building. With research on bioinformatics mushrooming, both theoretical and wet-lab based bioinformatics aided works are often deemed very procedural and paraphernalia that these are not easily accessible to those who want to use the "tools for biology". Additionally, the career-driven paths using bioinformatics is tacit by the fact that one needs to attend to earn programming skills which is not always the case. This book aims to be an interface between those who aim for bioinformatics and apply research with focus on Q and A on career growth. *A great saying goes "If you want more, you have to require more from yourself."* This also applies to bioinformatics. Happy reading!

Prashanth N. Suravajhala

Acknowledgments

I thank Aninda Bose and Chandra Shekhar of Springer who have supported me all through the making of the book. Although the cartoons and illustrations were ideated by me, full credits to Partha Paul for bringing life to them.

My sincere gratitude goes to my mentees without whose thoughts this book would not have been here today. Likewise, I owe appreciation to my wife Renuka and my daughters Bhavya and Nirmala who always stood by me.

My peers in Bioclues.org and bioinformatics.org, ex-colleagues and researchers in India, Denmark, US and Japan, countless "e-colleagues", also contributed to my discussions. I sincerely thank Cox Murray, Jeff Bizzaro, Madhan Mohan and Pawan Dhar who were generous enough to have responded to the questionnaire.

My grandparents—Shri D. S. Sastry and D. S. R. Murthy are always remembered with fond love and affection. They have helped me in imparting clarity, coherence and brevity, to the text.

Finally, the book would not have come into a good shape without the help of Springer reviewers, friends and well-wishers, but not the least the author sincerely thanks the Springer typesetting team, Messers Nalini Gyaneshwar, Kamiya Khatter et al for bringing the manuscript in shape.

Contents

Chapter 1
Whither Bioinformatics?

Ever since the word 'Theoretical Biology' was coined by Paulien Hogeweg in 1978, bioinformatics, the current word has steadfastly come into existence with many biologists taking a leaf out of this discipline. Researchers by now know that bioinformatics is a mere tool, whereas its sister concern, computational biology, is deemed as a discipline. With bioinformatics burgeoning in the late 1990s, we relate the commencement of data deluge to the animistic knowledge that bioinformatics has brought in, lessening the scale of experimentation. Authentic bioinformatics, however, will not gain significant interest for researchers, at least until the wet-laboratory biologists take a leap forward in acclimatizing the split half-term in bioinformatics. The figure of dogmas is pivotal in bringing the collaboration between biologists and cross-disciplinarians across biology as the event of dogmas in turn has introduced a plethora of new relationships between scientific studies and molecular biology. In effect, researchers have asked several questions on specialized mechanisms, if any that may be discovered in the advent of bioinformatical knowledge. This collaborative knowledge owes its impetus to the differentiation of an independent eccentric science, viz. systems biology (SB). So to ask whither bioinformatics into the enunciation and practice of the bioinformatical tools and scientific methods is a candid query.

Bioinformatics, since ages, has created a process of reasoning that was certainly not dependent on biology alone. Prior notions of intelligent algorithms clubbed with statisticians' skills, IT scientists' inclination, physicists' predictions, chemists' corner, and mathematicians' mind are a necessity to perform bioinformatics research. Not all disciplines can be made up by an individual alone but needs unicentric efforts to meet the goals to derive bioinformatical knowledge. For example, the next generation sequencing (NGS) technologies have enabled non-sanger based sequencing technologies with an unprecedented speed, thereby enabling novel biological applications. However, before bioinformatics and NGS stepped into the limelight, it must be noted that the NGS had overcome torpor in the field with the help of several cross-disciplinarians. It would never have been easy to stir up this understanding without the rapid involvement of the multi-faceted scientists who have transformed biology as a whole. This obviously has advantages of building up cross-disciplines, thereby deepening the crevasse

P. N. Suravajhala, *Your Passport to a Career in Bioinformatics*,
DOI: 10.1007/978-81-322-1163-1_1, © Springer India 2013

between eccentric biology and information science, the latter constantly teaming up with the former to signify its discoveries with dogmas.

The greatest challenge facing the molecular biology community today is to make sense of the wealth of data that have been produced by genome sequencing projects. Conventional biology research was deemed always to be in the laboratory, and so there were no mammothian data produced until the data deluge and explosion of genomic scale in the late 1990s. Thus, we are in an age of computing-to-research process. There are two different challenges one would pose: (1) sequence generation and (2) ensuing storage of the plethora of sequences generated in the laboratory with specific understanding and investigation using computers and artificial intelligence. That said, understanding the biology of an organism is a trivial issue as there are a number of focused research areas at different levels of 'omics'-es, viz. genomics, proteomics, functomics, transcriptomics, need to being carried out at different levels. One of the foremost challenges today is to ensure that such data are efficiently stored, used through three forms of Es—extracting, envisaging, and elucidating this mass of data. A meaningful interpretation of such data must be done before one analyzes the complete volume for interpreting it or what we call 'annotating' manually. In conclusion, discerning the function using computer tools must be the focus so as to have meaningful biological information explained.

The journey of transcriptomics starts with the discovery of ribonucleic acids in 1869 (http://transcriptomics.net) followed by their role in protein synthesis and as catalyst in various biochemical reactions. However, the term transcriptomics first appeared in 1998 in the scientific literature (http://www.nature.com/omics/subjects/transcriptomics/1998.html) concurrent with different 'omics' terminologies. Different 'omes' and their respective descriptions are summarized in Table 1.1.

Table 1.1 Components defining different 'omics' technologies. The word 'ome' refers to 'many' or 'monies.' For example, genomes indicate the study of many genes

'Omes'	Description
Genome	The full complement of genetic information both coding and noncoding in an organism
Proteome	The complete set of proteins expressed by the genome in an organism
Transcriptome	The population of mRNA transcripts in the cell, weighted by their expression levels as transcripts copy number
Metabolome	The quantitative complement of all the small molecules present in a cell in a specific physiological state
Interactome	Product of interactions between all macromolecules in a cell
Phenome	Qualitative identification of the form and function derived from genes, but lacking a quantitative, integrative definition
Glycome	The population of carbohydrate molecules in the cell
Translatome	The population of mRNA transcripts in the cell, weighted by their expression levels as protein products
Regulome	Genome wide regulatory network of the cell
Operome	The characterization of proteins with unknown biological function
Synthetome	The population of the synthetic gene products
Hypothome	Interactome of hypothetical proteins

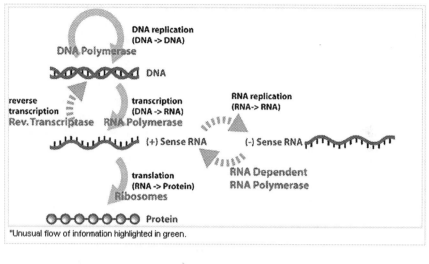

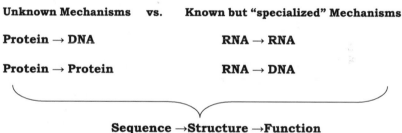

Fig. 1.1 An overview of the dogma of molecular biology with known specialized and unknown mechanisms/flows. (Image courtesy: Daniel Horspool)

Why is bioinformatics interesting? All the central biological processes revolve around bioinformatics tools that need to be developed, which possibly leeway in understanding the sequence–structure–function relationship (See Fig. 1.1). *DNA sequence determines protein sequence which determines structure and function.* Why is it that we end up with protein as a determiner for every analysis? The simplest answer is we would have less noise when we deal with protein sequences wherein we deal with 20 odd amino acids to narrate upon results unlike the several compositions of four bases, viz. ATGC compendium of six reading frames translating into amino acids. This integration of information making up biological processes would allow us to understand the complete repertoire of the biology of organisms. However, the challenge faced by the biology community, especially on the inordinate data, is more from the umpteen genome sequencing projects. Traditionally, wet-laboratory biologists carry experimental work even as the huge increase in the scale of data being produced from time to time could be better facilitated by *in silico* research process. With the help of high-performance computing (HPC), sequences generated can be sporadic and further analyzed. Nevertheless, given the fact that molecular biology of a system is very complex, understanding and disseminating the information is to be carried out at different levels using the 'omes' including the genome, proteome, transcriptome, and metabolome levels. There is a need for researchers, especially from the wet-laboratory community, to herald bioinformatics indefatigably both in academia and industry.

What discoveries interest researchers? Looking at the dogmas, it is still not clear whether or not a protein can replicate another protein or a DNA can be obtained by direct reverse translation. It would be intriguing to understand if these could really happen in an organism. Can a genomic repertoire take shape in understanding the dogmas bottom-up? There is an aspiration but puny hope that bioinformatics can handle this. For example, protein–protein interactions (PPI) play a huge role in understanding the function of proteins. Various bioinformatic tools have been developed that allow researchers to compare proteins. Such comparative studies using algorithms such as BLAST (Altschul et al. 1990, 2005; Alstchul 1991) and other tools were carried out to distinguish unique proteins from paralogs which later might have resulted from gene duplication events. The genomes sequenced so far were helpful in predicting not only evolutionary relationships but also identified function for the genes through functional genomics (Link et al. 1997). The *in silico* methods such as homology search, presence of motifs, domains and signature sequences, orthology mapping, and radiation hybrid transcript mapping (Avner et al. 2001) are available for descriptive predictions of proteins with known (and sometimes unknown) function. However, these employed methods possibly might have a lot of false positives unless *in vitro* and/ or *in vivo* experiments are followed to validate them. Moreover, these methods do not reveal a predicted function of hypothetical proteins (HP), thus making predictions more insignificant. Although all these methods are being employed by researchers, screening of HPs for novel translatable candidates is not often used and the researcher repeatedly performs the screening with laborious wet-laboratory experiments. Furthermore, the proteins whose function remains unknown (i.e., those that remain hypothetical) and that are targeted to different organelles, especially mitochondria, could be important.

Many protein sequences contain motifs or short signature sequences called equivalogs (Haft et al. 2003), which are conserved in several organisms. These are a set of homologous proteins conserved since their last common ancestor with respect to function (Lee et al. 2002). Some of these proteins might have a chance to be duplicated in organisms. It is therefore necessary to understand the genomic context of such proteins. An example of equivalog model is TIGR00658, identified as ornithine carbamoyltransferase. However, this enzyme is also known to act in an arginine biosynthesis pathway from ornithine (TIGR00032 and TIGR00838) in *Yersinia pestis* and arginine degradation (TIGR00746 and TIGR01078) in *Streptococcus pneumoniae*. The TIGRFAMs models, a TIGR family database, include equivalog models that have been used extensively in genome annotation. In addition, proteins with weak sequence similarity and no relevant structural homologies usually do not have known cellular function; such proteins are discarded from well-known proteins. When annotating proteins, new molecular role for known cellular function is carefully addressed and curated. In many cases, there are numerous proteins which fall under domains whose functions are essentially known but they have no genuine role played in genomes. In addition, when annotating, if a reference genome is considered besides comparing sequences in UniGene database (see web references) with selected protein reference sequences, the alignments would possibly suggest function of a gene and finally the possibility of annotating the protein as hypothetical would be reduced. For example, many proteins in humans have been named as some repeat domains (for example, accession #CAB98209.1) maintaining homology to some known domains and all of them fall under a large category of domains. This does not necessarily mean that all these proteins make up a function. There are also some instances of some proteins already similar to some organisms not showing up the function even though possibly studied from *in silico* and a few wet-laboratory studies. Two such examples, one from eukaryotes and the other from bacteria, are discussed in what follows:

1. The Ankyrin repeat domain 16 (Ankrd16) has protein similarities in mice with 100 % (Accession #NP_796242.1), humans with 85.7 % (Accession #NP_061919.1), Xenopus with 71.1 % (Accession #NP_001088685.1), and *Danio rerio* with 66.6 % (Accession # NP_001017563.2). This annotation as revealed by GenBank and UniGene reference (http://www.ncbi.nlm.nih.gov/UniGene/clust.cgi?ORG=Mm&CID=260201), might have an update at a later point of time, when new orthologs as identified from other metazoan sequences keep adding up to the annotation.
2. In bacteria, the proline proline and glutamic acid (PPE) and the proline glutamic acid (PE) gene families comprise many unique genes, some of them novel and labeled as hypothetical. Of them, many are known to be pseudogenes (Marri et al. 2006). The 10 % of the coding DNA of *Mycobacterium tuberculosis* constitutes PE and PPE family genes and is involved in gene expression upon infection of macrophages, some of them as antigens mediating role in pathogenesis or in virulence. These were characterized while the expression levels and the functions of select PE/PPE family genes during various phases of infection (latent/mild/hypoxic) with *Mycobacterium tuberculosis* (Kim et al. 2008) were studied (Fig. 1.2).

The aforementioned examples discussed are all a resultant of explosion of bioinformatics tools during the last three decades. Have bioinformatics technologies revolutionized genomics and proteomics? Well, there has been a focus on molecular medicine which paved the way for establishing intervention and treatment of well-known diseases to proactive prediction and prevention of disease risk. These approaches should really require new informatics systems that will link large-scale databanks and special programs for data mining and retrieval in bioinformatics and chemoinformatics. All the wet-laboratories should be able to provide a platform for powerful new molecular diagnostic tools along with multianalyte assays for expression of genes and proteins in different patterns of diseases. With researchers scaling the ladder of bioinformatic progress by leaps and

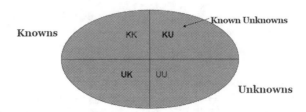

Fig. 1.2 The importance of Known Unknowns aliased 'hypothetical genes' in the genome, illustrated in the form of a checker board. The Known is acronymed '*K*' while the Unknown '*U*'. Apparently, we seldom find '*UU*'s as it is a misnomer here. Unless the genome is sequenced, we find genes evaluating and devaluating

Table 1.2 Pros and cons of different methods in annotating sequences

1. Sequence-based methods
Pros: most known/reliable method
Cons: BLAST hits are electronically annotated and turn out to be false positives
2. Structure-based methods
Pros: based on active site characterization/global fold similarity
Cons: free energy minima always need to be set/obligation
3. Associated-based methods
Pros: based on domain is/phylogenetic profiles
Cons: lack of conserved proximity does not indicate lack of functional association
4. Proteomics-based methods
Pros: based on protein interactions domains. Gaps or holes in known pathway can be assigned. Function awaits a protein to be characterized
Cons: lots of false positives

bounds, there is a need for enhanced understanding of the interactions in a system (organism). What are the components that interact with each other? What is the outcome of such interactions? Do interactions alone provide us the functional decipherment? Should we just be sufficed with the progress made on say, cures for diseases by the year 2050? Should we reach consensus on the combination of tools, viz. rapid and inexpensive DNA sequencing technologies, HapMap project, dollar one genome (DOG), etc.? This we hope will let us understand precisely how bioinformatics transits from research, to vocation and avocation (Table 1.2).

1.1 Bioinformatics 'Aging' in Systems Biology

Systems biology has gained a lot of excitement these days. Of late, biologists have been actively engaged in this discipline in different forms when molecular biology merged with multi context disciplines. During this process, SB ran into several definitions. To answer what is a system: We could think of multiple organelles existing in our human body as we use components to describe entities in a system.

Table 1.3 Timeline eventing important spheres in bioinformatics

- 1859 Charles Darwin's 'origin of species'
- 1944 Avery, MacLeod, McCarty: DNA is the genetic material
- 1953 Structure of DNA
- 1955 Complete sequencing of insulin
- 1988 National Center for Biotechnology Information (NCBI) founded
- 1988 Sanger Centre, Hinxton, UK
- 1994 EMBL European Bioinformatics Institute, Hinxton, UK
- 1995 First bacterial genomes completely sequenced
- 1996 Yeast genome completely sequenced
- 1999 Fly genome completely sequenced
- 2000 bioinformatics.org and open source
- 2001 Human genome and bioinformatics ****Systems Biology***
- 2002–2004 umpteen genomes sequenced
- 2005 EVOLUTION in terms of bioinformatics as a breakthrough
- 2007 Personal genomics ~ ~ person 'omics'
- 2008 The hypothetical proteins and orphan genes???
- 2011 Predictive biology approaches

As we integrate various vehicular components to construct a vehicle, we describe components such as organelles to make up a living system. The biology of system is called systems biology. Every system has an effect on its environment and so are the components in a system, even as the components entitled to SB include genes, proteins, metabolites, and enzymes as minor entities, while cells, tissues, organelles, and organs as major components. Hence, interactions among the components would be interesting to value SB. While a system could have many organelles and the components that make up the flow of a system, they are bound to interact with one another. For example, enzymes, proteins, metabolites, genes, DNA, and functional protein domains are known to interact with each other. Integrating all the interactions of components indicates: which survives (and competes) the best while the ultimate goal of SB is to exploit the interplay among the components. From a reductionism's point of view, researchers define SB based on whether the components in a system are interacting with each other, mutations arising and falling, proteins evaluating and devaluating, strains adapting and unfitting in the environment, and some genes if lost and found (Table 1.3).

1.2 Defining Systems Biology Through Omics: The Two Paradigms

Is systems biology all about the genes making up the proteins and how the components processing in a system interact with each other? The fields of omics in the recent past have believably revolutionized biomedicine and by far means there needs to be a focus on change in defining these upcoming omics-es. Huang S's classification of SB has yielded the loose and the apparent but broadened definitions from dynamics and reductions approach (Huang 2004). The dynamicity of SB is based on a pure level where the system is based on models and networks: be it quantitative or qualitative, whereas the reductionism defines SB based on the high-throughput methods involving different molecular biology techniques. Overall, the loose definition applies to projects exploring individual biological networks, while the broadened but still 'derivative' definition is the outgrowth of theoretical models along with systems theory across inter-disciplinary sciences such as engineering, mathematics, statistics, artificial intelligence, and so forth. However, many authors (Tracy 2008; Cornish-Bowden et al. 2007; Huang and Wikswo 2006; Strömbäck 2006; Bruggeman and Westerhoff 2007) have recently deliberated that the concept of the gene resulting in omics has begun to outlive its usefulness while they felt that the SB could be projected into several dimensions keeping in view the multifaceted systems' complexity of living organisms. With SB maturing, researchers have started proposing an alternative means to define gene based on a richer explanation: Genetic functor, or genitor, a sweeping extension of the classical genotype/phenotype paradigm that describes the 'functional' gene (Fox Keller and Harel 2007). Thus, we could understand the dynamic behaviors of molecular associations implicitly known from various methods and technologies integrating one or more of the SB data:

Overall, SB can be envisaged keeping in view the following points:

1. Systems biology is conceptualized in terms of PPI. The interplay between components in systems is exploited between protein–protein, domain–domain, DNA–DNA as a whole or even a protein–DNA.
2. The interactions among the components are better explained in such a way that what is in theory need not fit practically implicating that a hypothesis-driven approach need not always be experimental (biological) driven.
3. With some answers to questions like if there are interactions known, we can take measure of unknown interactions in a system, SB approaches toward understanding *bona fide* PPI.

Does SB back biologists? There are specific traits that make up PPI networks: Everything in biology is better explained through interactions while the interactions are a priority in accordance with organization, cooperability, and mapping the components in a system. The SB signifies if components interact with each other. This led to the birth of several disciplines such as systems molecular medicine, immunological SB; local and global metabolic profiling, systems diagnostic therapy, and systems drug development, all budding across nascent biology disciplines. Although the PPI are outcomes of almost all cellular processes, there is diversity in protein interactions, i.e., all proteins share common properties at a certainty. For example, the distortion of protein interfaces lead to development of many diseases and to understand its mechanism, we lead to PPI experiments. When proteins recognize specific targets and bind them, it results in conservation that depends on structural and physicochemical properties. The nature and applications of SB with respect to PPI were well reviewed elsewhere (Huang 2004; Tracy 2008; Cornish-Bowden et al. 2007; Huang and Wikswo 2006; Strömbäck 2006).

1.3 Is Biology Explained Through Protein–Protein Interaction Networks Alone?

Apart from the three most common omics-es, viz. 'Gen-omics,' 'Prote-omics,' and 'Transcript-omics,' bioinformatics and biology researchers have been taking up omes and omics-es very rapidly as is evident from the use of the terms in PubMed (Dell et al. 1996). As a result, a variety of omics disciplines such as phenomics (Schork 1997), physiomics (Chotani et al. 2000; Gomase and Tagore 2008), metabolomics (Kuiper et al. 2001; Fiehn 2002), lipidomics (Han and Gross 2003), glycomics (Gronow and Brade 2001), interactomics (Govorun and Archakov 2002), cellomics (Taylor et al. 2001) have begun to emerge, each with their own set of instruments, techniques, reagents, and software. These have driven new areas of research consisting of DNA and protein microarrays, mass spectrometry, and a number of other instruments that enable high-throughput analyses.

While genomics forms a main hierarchy of classification, there are many other omics-es which fall under a clad of primary (gen) omics' enabled SB, e.g., functional genomics, comparative genomics, computational genomics, and phylogenomics. With over 1800 microbial genomes sequenced or being sequenced today and the number still increasing, another set of omics called metagenomics aims to access the genomic potential of an environmental sample. It would answer to some of the questions we posed in the earlier sections. This environmental 'omics' bridges integration of metagenomics with complementary approaches in microbial ecology (Schloss and Handelsman 2003).

While the mapping of PPI is a key to understand biological processes through interactomics, many technologies have been reported to map interactions, widely applied in yeast. At present, the number of reported yeast protein interactions truly validated by at least one other approach is low with the amount of throughput it takes to process (Cornell et al. 2004). This is because of the false discovery rate of proteins interacting with their partners. With the advent of virtual interactions, the growth of false positives also increased, thereby allowing the researchers to keep a track of finding these false positives through statistical inference. Any dataset of interaction map is complex while tools to decipher true positives are being developed in the form of mark up languages such as system biology markup language (SBML) (Hucka et al. 2004). The mapping of human–protein interaction networks is even more complicated, suggesting that it is unreasonable to try mapping the human interactome; instead, interaction mapping in human cell lines should be focused along the lines of diseases or changes that can be associated with specific cells (Figeys 2004). This '-omics revolution' would force us to

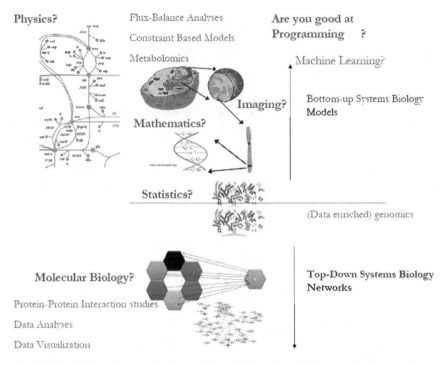

Physics?

Flux-Balance Analyses

Constraint Based Models

Metabolomics

Are you good at
Programming   ?

Machine Learning?

Imaging?

Mathematics?

Bottom-up Systems Biology
Models

Statistics?

(Data enriched) genomics

Molecular Biology?

Protein-Protein Interaction studies

Data Analyses

Data Visualization

Top-Down Systems Biology
Networks

Fig. 1.3 Quantitative picture of various omics and the various fields, an enthusiast can take up

re-evaluate our ability to acquire, measure, and handle large datasets. The omic platforms such as expression arrays, MS, and other high-throughput methods have enabled quantization of proteins and metabolites derived from complex tissues. Applying SB, the integrated analysis of genetic, genomic, protein, metabolite, cellular, and pathway events are in flux and interdependent. With the onset of various datasets, it necessitated the use of a variety of analytic platforms as well as biostatistics, bioinformatics, data integration, computational biology, modeling, and knowledge assembly protocols. Such sophisticated analyses would definitely provide new insight into the understanding of disease processes through phenome–genome networks and interactomic studies (Lage et al. 2007). In this regard, SB clubbed with interactomics, more appropriately considered as a process containing a series of modules, aims to provide tools and capabilities to carry out a wide range of tasks (Morel et al. 2004). Even as protein analysis is known as a field of research with a long history, several developments of a series of proteomics approaches including MS opened the door for a synergistic combination with genomic sequence analysis, focusing on aspects of genomewide transcription control, regulomics. In analogy with all the other omics-es, a combination of MS-based proteomics with *in silico* regulomics analyses can produce synergistic effects in the quest to understand how cells function (Werner 2004). Carrying this further, it has been suggested that the term 'translatome' could be used to describe the members of the proteome weighted by their abundance, and the 'functome' to describe all the functions carried out by them (Greenbaum et al. 2001). However, there are still many difficulties resulting from the disorderliness and complexity of the information. To overcome this, removing noisy data and finding false positives could be enhanced using various tools to some degree. However, these can also be overcome by averaging broad proteomic categories such as those implicit in functional and structural classifications (Fig. 1.3).

1.4 Systems Biology in Wet-Laboratory

Fundamental biological processes can now be studied by applying the full range of omics technologies (genomics, transcriptomics, proteomics, metabolomics, etc.) using the same biological sample and high-throughput methods such as MS (McGuire et al. 2008; Kim et al. 2008). Wide array of assays including high-throughput methods such as tandem mass spectrometry (MS/MS), Yeast two hybrids (Y2H), and pull-down assays are preferentially used to navigate them. Clearly, it would be desirable if the concept of sample were shared among technologies such as MS for that, until the time a biological sample is prepared for use in a specific omics assay, its description is inherently technology independent. However, the compulsion for accurate analyses of all these high-throughput methods is to remove redundant and false-positive data. Redundancy of data has been the biggest threat for causing errors in data usage. Sharing a common informatics' representation would encourage data sharing, leading to decrease in redundant data and the potential for error. The recent introduction of *WikiProteins* has been a worthy effort that brought all annotators to come together on a common platform (Mons et al. 2008). This would result in a significant degree of harmonization across different omics data standardization activities, a task that is critical if we are to integrate data from these different data sources (Morrison et al. 2006). The bioinformatics applied to omics' are varied and particularly noteworthy or characteristic of proteomics research, e.g., 2DE analysis or MS. Another important task of bioinformatics is the prediction of functional properties through ontology-based functional networks from a vast amount of databases.

Apart from the above-discussed issues, genome technologies are being carried out in every major model system. For example, new technologies are being developed to rapidly identify mutations or small molecules that increase the life span for aging-related research. While the DOG recently has been known to play a role as a model system for cancer, because of its similarities to human anatomy and physiology, it may prove invaluable in research and development on cancer drugs (Khanna 2006). Inversely, as dogs too naturally develop cancers they may share many characteristics with human malignancies. This probably would accelerate genomewide, cross-comparison of organisms for finding function of more genes ultimately using drug discovery development.

1.5 Metabolomics

Metabolomics has come into sight as one of the newest 'omics' science with dynamic portrait of the metabolic status of living systems. The analysis of the metabolome is particularly challenging as it has its roots in early metabolite profiling studies but is now a rapidly expanding area of scientific research in its own right. It is a science employed toward the understanding of global SB (Rochfort 2005). The metabolomic tools aim to fill the gap between genotype and phenotype permitting simultaneous monitoring molecules in a living system. The smartness of using metabolic information could be applied in translating into diagnostic tests as they might have the potential to impact on clinical practice, and might lead to the supplementation of traditional biomarkers of cellular integrity, cell and tissue homeostasis, and morphological alterations that result from cell damage or death (Claudino et al. 2007). Metabolomics has been widely applied to optimize microorganisms for white biotechnology even as it spreads to the investigation of biotransformation and cell culture. Together with the other more established omics technologies, metabolomics aims to contribute in different spheres ranging from understanding of the *in vivo* function of gene products to the simulation of the whole cell in the SB approach. This will allow the construction of designer organisms and yet another science synthetic biology evolves (Oldiges et al. 2007). Although metabolomics measures multiparametric response of living systems to genetic modification, there is a consistent debate of synonymy with metabolomics. Admittedly, there is a concurrence of the former being associated with NMR while the latter being associated with mass spectroscopy. This part of microbial transformation has led several standards for these two meta-omics' delivering SB tools (Oliver Fiehn et al. 2006).

1.6 Mitochondriomics

Mitochondria are semiautonomous organelles, presumed to be the evolutionary product of a symbiosis between a eukaryote and a prokaryote. The organelle is present in almost all eukaryotic cells in an extent from 10^3 to 10^4 copies. The main function of mitochondria is production of ATP by oxidative phosphorylation and its involvement in apoptosis. The organelles contain almost exclusively maternally inherited mtDNA, and they have specific systems for transcription, translation, and replication of mtDNA. Mitochondrial dysfunction has been correlated with mitochondrial diseases where the clinical pathologies are believed to include infertility, diabetes, blindness, deafness, stroke, migraine, and heart-, kidney-, and liver diseases (Reichert and Neupert 2004).

Recently, cancer was added to this list when investigations into human cancer cells from breast, bladder, neck, and lung revealed a high occurrence of mutations in mtDNA. With the understanding of the role of mitochondria in a vast array of pathologies, research on mitochondria and mitochondrial dysfunction has in the last decade yielded a huge amount of data in the form of publications and databases. Yet, the field of mitochondrial research is still far from exhaustion with many essentials waiting to be discovered. The recent identification of a number of proteins targeting mitochondria has enabled immense interest to understand the function of some genes unnoticed in mitochondrion (Calvo et al. 2006). With only 13 proteins sitting inside mitochondria through oxidative phosphorylation, and over 1,500 estimated proteins targeting this tiny organelle, identifying complete protein repertoire in this machinery could decipher the biology behind mitochondria or what makes us breathe. A complete set of mitochondrial proteome syntenic with other eukaryotes has just started and there is a promise in understanding how the organelle proteomes and interactomes could essentially be used to develop into SB (Calvo et al. 2006).

1.7 'Omic' Challenges in Systems Biology

Bioinformatics has enabled all-against-all comparison distinguishing unique proteins from proteins that are paralogs resulting from gene duplication events. The last two decades have seen an avalanche in databases while algorithms such as BLAST allowed such comparisons. In post-genomic era, the genomes sequenced so far would essentially cover the future of omics in them as they enable predicting not only evolutionary relationships but also make use of different approaches used in identifying function of genes. This functional genomics is the cause of understanding how proteins interact with each other and network in the living organism. The gene or protein function could be ascertained based on physiological characterization or if the two proteins are known to be physically interacting with each other or virtually interacting with each other. The SB approaches in present-day bioinformatics have brought in a special emphasis on association-based networks in the form of virtual interactions, thereby making up the possibility of phenome–genome networks grow bigger (Lage et al. 2007). Ultimately, it makes sense when such interactions bring out a function and find a candidate for a disease. The increase in GenBank accessions resulted not only in number of genes identified but also the number of citations these accessions refer to. While various databases and terms have been defined, several omics-es are reported from time to time at http://www.omics.org (Fig. 1.4).

The last 10 years have not only seen the rise of bioinformatics producing an unprecedented amount of genome-scale data from many organisms but also the wet-laboratory research community has been successful in exploring these data on using bioinformatics many challenges still persist. One of them is the effective integration of datasets directly into approaches based on mathematical modeling of biological systems. This is where SB has bud resulting in top–down and bottom–up approaches. The advent of functional genomics has enabled the molecular biosciences to come a long way toward characterizing the molecular constituents

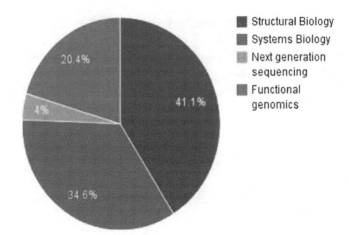

Fig. 1.4 Proportion of research articles published in different omics-es in PubMed as on September 1, 2012

of life. Yet, the challenge for biology overall is to understand how organisms function. By discovering how function arises in dynamic interactions, SB is everywhere addressing the missing links between molecules and physiology. Top–down SB identifies molecular interaction networks on the basis of correlated molecular behavior observed in genomewide 'omics' studies. On the other hand, bottom–up SB examines the mechanisms through which functional properties arise in the interactions of known components. Applications in cancer are a good example to counteract these two major types of complementary strategies (Stransky et al 2007). Several web-based repositories have been established to store protein and peptide identifications derived from MS data, and a similar number of peptide identification software pipelines and workflows have emerged to deliver identifications to these repositories. Integrated data analysis is introduced as the intermediate level of a SB approach and as a supplementary to bioinformatics to analyze different 'omics' datasets, i.e., genomewide measurements of transcripts, protein levels or PPI, and metabolite levels aiming at generating a coherent understanding of biological function (Steinfath et al. 2007). Furthermore, existing and potential problems/solutions such as de facto experimental and the following bioinformatical challenges might hold prospective in the near future:

1. Challenges in high-dimensional biology (HDB): Recently, the term HDB has been proposed for investigations involving high-throughput data (Mehta et al. 2006). The HDB includes whole-genome sequences, expression levels of genes, protein abundance measurements, and other permutations. The identification of biomarkers, effects of mutations, and effects of drug treatments and the investigation of diseases as multifactor phenomena can now be accomplished on an unprecedented scale.

2. Finding the function of HP: Another feature of PPI maps is to find the function of unknown proteins. PPI has become a very common step in annotation of a protein. Various tools such as iHOP (http://www.ihop-net.org), STRING (http://string.embl.de), etc. aid the researchers to find if there are interacting partners of protein of interest. The data could be visualized through tools such as Cytoscape (www.cytoscape.org), VisANt (http://www.visant.bu.edu), and Osprey (http://biodata.mshri.on.ca/osprey/servlet/Index), etc. for further analyses. The nearest partners would essentially mean that the hypothetical or uncharacterized protein could play a function similar to its interactor(s). In the context of PPI networks, we could consider if a model is to be developed from the network or a network is to be generated with an already established model. Precisely, the putative function of a protein could be better known from a PPI network to develop a model from it. Information on 'known' or 'unknown' PPI is still mostly limited but integrating tools such as these could generalize a way to find *bona fide* function.

1.8 Are Interactions Based on the Nature of Binding?

Does close homology between two proteins confer that they do interact in the same manner? Yes, they do and confer evolutionary constraints in lieu of structural divergence while remotely related proteins have a different interaction mode (Drummond et al. 2005). Also, conservation of protein interface indicates the average conservation of the rest of the protein. While all these form an integral part of SB, apart from the novel interactions that arise based on the type of homology, there are interactions based on the binding entity, viz. stable and transient. The former interactions are consistent and bookmarked while the latter are temporary. There are interacting proteins that might co-express indicating that the expressed proteins, which evolve slowly are normalized wherein the normalized difference between the absolute expression data is calculated based on several tools such as microarrays (Drummond et al. 2005). However, there are other techniques such as density gradient and virtual pull down assay methods cited as above beginning to be understood and substantiate above views.

As thousands of new genes are identified in genomics efforts, the rush is on to learn something about the functional roles of the proteins encoded by those genes. Clues to protein functions, activation states and PPI have been revealed in focused studies of protein localization. A meta analysis of data derived from genomewide studies of aging in simple eukaryotes will allow the identification of conserved determinants of longevity that can be tested in other mammals (Khanna 2006; Kaeberlein 2004). Adding to the various high-throughput methods, technical breakthroughs such as GFP protein tagging and recombinase clones, large-scale screens of protein localization are now being undertaken to understand the function of the proteins (O'Rourke et al. 2005).

1.9 Fundamental and Best Practiced Tools for Annotating Proteins and Genes

In the recent past, various bioinformatics tools have been developed that allow researchers to compare genomic and proteomic repertoire. Comparative studies using algorithms such as Blast and databases are carried out to distinguish unique proteins from paralogs which later might have resulted from gene duplication events. The genomes sequenced so far were helpful in predicting not only evolutionary relationships but also identified function for the genes through functional genomics. Although many methods are being employed by researchers, screening of proteins for novel translatable candidates is not often used and the researcher repeatedly performs the screening with laborious wet-laboratory experiments. To increase the sensitivity, further clues on tissues, and development stages from the queried gene's sequences could be surveyed using tools such as gene expression omnibus (GEO) or UniGene-EST or cDNA profile database. Furthermore, protein link to genomic location specified by transcript mapping, radiation hybrid mapping, genetic mapping, or cytogenetic mapping as available from GenBank resources would improve the understanding on protein annotation. Besides this, whether or not a protein contains a polyadenylation signal could be an added knowledge to meet the criteria of well-annotated proteins. This is because tools such as MEME reveal many 3′ UTRs forming conserved motifs, which indicates these regions appear more conserved than expected. This means, higher the conservation, greater is the duplications and greater is the chance of being not annotated or 'hypothetical'. There seems to be many unique genes which are overrepresented in the form of duplications; a simple search in GenBank gene list would reveal that there are several accessions duplicated. For example, in case of the gene FusA2a, bona fide accession is mapped to CAD92986 and yet a few of the isoforms/unique genes remain unknown (For example, CAD93127). In summary, there could be many proteins less annotated and yet many tools are known to describe the function. This leaves to beg a question, what would be the fate of proteins that cannot be annotated through some tools or in contrast how many best tools amongst them are used to describe or annotate a protein?

Apart from BLAST and FASTA, the sequence-based feature annotation is applied by RefSeq using several tools, viz. BEAUTY X-Blast Enhanced Alignment Utility and PROSITE. While many other variants of BLAST including PSI Blast and PHI Blast, sequence alignments using ClustalW, ClustalX, and Cobalt are used, not all the tools are used in tandem to eliminate false positives. Whether the protein is soluble or insoluble is known through TopPred; the topology of protein with the orientation and location of transmembrane helices attribute to the function. Additionally, orthology mapping using tools such as HomoMINT are used which increases the chance of the protein annotation. With the central dogma beyond the age today in bioinformatics, viz. sequence specifies structure and function; annotations have become mightier to further manually curate allowing researchers to perform experimental analyses for some proteins. The structures of proteins not only provide functions but the shapes exhibited by the proteins allow them to interact selectively with other proteins or molecules. This specificity is the key for the proteins to interact with another protein thereby inferring the function. However, most of the bioinformatical analyses are misleading unless biochemical characterization is carried out. Further, the protein annotation has gained much importance with the introduction of many metazoan genome sequencing projects in addition to the 1,000 genomes project that is in progress. With 40–50 % of identified genes corresponding to proteins of unknown function, the functional-structural annotation screening technology using NMR (FAST-NMR) has been developed to assign a biological function which is based on the principle that a biological function can be described based on the basic dogma of biochemistry that the proteins with similar functions will have similar active sites and exhibit similar ligand-binding interactions, even though there is a global difference in sequence and structure. Tools such as combinatorial extension which confer structure similarity, DALI for NMR, finally determining function, PvSOAR, and Profunc—given a 3D structure, aims at identifying a protein's function have been widely used. However, there are many other methods such as the Rosetta Stone method, phylogenetic profiling method, and conserved gene neighbors that have been widely employed and being accepted by the scientific community.

Biological function of proteins would help in the identification of novel drug targets and helps reduce the extensive cost of practical examinations on several candidates. With the enormous amount of sequence and structure information availability, innumerable automated annotation tools for proteins have also been generated. One such example is automated protein annotation tool (APAT), which uses markup language concept to provide wrappers for several kinds of protein annotations. While FFPred is available to predict molecular function for orphan and unannotated protein sequences, the method has been optimized for performance using a protein feature-based method through support vector machines (SVMs) that does not require prior identification of protein sequence homologs. It works on the premise of post-translational modifications, Gene Ontology, and localization features of proteins. Yet another tool, viz. VICMpred, aids in broad functional classification of proteins of bacteria into virulence factors, information molecule, cellular process, and metabolism molecule. The VICMpred server uses an SVM-based method having patterns, aminoacid, and dipeptide composition of bacterial protein sequences. ConSeq and ConSurf have been widely applied in predicting functional/structural sites in a protein using conservation and hypervariation.

The final part of annotation can be studied through interactions and associations. All interactions are associations, while not all associations are interactions. The association tools, viz. search tool for the retrieval of interacting genes/proteins (STRING), GeneCards, IntAct, MINT, biomolecular interaction network database (BIND) which have been enhanced as biomolecular object network database (BOND). With BIND inside, BOND is a comprehensive database that helps in the annotation of proteins through unique object-based interaction studies. Although there are other variants of some of these databases such as GeneAnnot and

Table 1.4 Best practiced tools for myriad of protein functional annotations

Tools/Servers	Interaction/association	Output	Comments on methods	Annotation
Blast/FASTA	Protein sequence database	Close and distant candidates	Heuristics	Homology
Pfam/GO	Protein sequences	Ontology based	Pattern based	Ontology
VICMPred	Annotated protein sequences	Functional information	Machine learning based	Functional annotation
Interpro/Prodom	Protein sequence motifs and domains	Protein family and domains	Domain based	Structural and functional annotation
MEME	Protein sequence motifs	Protein motifs	Statistical	Motifs identification and analysis
TopPred		Protein solubility conditions	Machine Learning based	Solubility/insolubility of proteins
Profunc	3D structure	Various functions	Machine learning based	Functional annotation
STRING/IntAct	PPI and database	Interactors	Pattern and mining based	Protein–protein interactions
TargetP/PTarget	Annotated proteins	Inter and intra cellular signals	Quantitative analytical	Signal sorting
APAT	Proteins	Myriad features	Markup language based	Miscellaneous
FFPred/RIGOR	Information based	Structural and functional element information	Structural elements based annotation	Structural and functional annotation
ConSeq/conSurf	Protein sequence and 3D structure	MSA, phylogenetic tree, various statistical scores, conserved residues on sequence, and structure of proteins	Sequence and structural evolutionary conservation and hyper variation	Structural, functional, and evolutionary annotation
MINT/BIND/BOND/GeneCards	Protein–protein interactions	Interactors and annotated pathways	Miscellaneous	Protein–protein interactions and pathways

GeneDecks of GeneCards, most of them are used for finding genes based on different queries.

Methods for predicting protein antigenic determinants from amino acid sequences were a crucial point for segment-level annotation of proteins. Since then, so many computational methods have been developed based on such basic and fundamental methods and pinpoint the importance of basic methods in the area of computational biology. Developed methods are being assorted in applications from sequence-based antigenic determinants to surface-based consensus scoring matrix approach for antigenic epitopes. Such developments significantly contributed to the refinement of existing and development of new and versatile techniques; but have roots in indispensable and conventional approaches.

Incorporating a systematic representation of fundamental and best practiced tools/servers to facilitate users for information would be useful even as additional features with their respective inputs, outputs, mode of action, and level of annotation have also been compiled (*See* Table 1.4), and will help experienced as well as beginners in the area of protein annotation.

1.10 Can Bioinformatics Influence Animal Experimentation?

Decades ago, legislation on the use of animals was enacted in many countries involving three R's': reduction, refinement, and replacement of animal models. Ever since this was enacted, there was a sudden buzz about laboratory animals and their use to be reduced, refined, and replaced wherever possible, for ethical and scientific reasons. The three R's concept was put forward by W.M.S. Russell and R.L. Burch in 1959 in The Principles of Humane Experimental Technique. A great detail on the three R's was reviewed by many in the interest of good and humane science. The word 'alternatives' came into use after publication of the book 'Alternatives to animal experiments' by David Smyth, in 1978.

With the arrival of bioinformatics and SB, the impact on animal experiments was slowly felt. The generation of high-throughput data in the form of genomics, transcriptomics, and metabolomics, biology has essentially transformed into a computational problem. Due to this reason, we believe that the role of computation in biology leading to reducing, refining, and replacing animal experiments needs to be reviewed and discussed. Let us review this question using two approaches.

1. **Reductionist approach**: Today the fields of omics have revolutionized fundamental biology and biomedicine. Greater attention needs to be paid on defining upcoming omics-es based on the three Rs. We believe that the first two R's—reduction and refinement—aptly fit into the category of definition where we may not completely replace animal experimentation but at least lessen the scale and usage of laboratory animals. Thanks to high-throughput techniques

through which we are able to better explore *in vitro* methods. However, the reductionist approach does not completely reduce or refine this process as this implies for smart experimentalists with humane touch. Homogeneity and environmental conditions play a major role in reducing the experimentation process. Greater the use of genetic homogeneity, greater is the chance to reduce the use of animal models. Similarly, greater the chance of maintaining and ensuring the conditions of the experiment more is the chance to reduce animal experimentation.

2. **Dynamic or a vibrant approach**: This applies to *in silico* models. Many computational models in biology are used nowadays. Dynamicists might not even go to that extent but think of plan b: considering a scale of sentience. A common question often asked is why not using animals that are small at the scale of taxonomy? But as computational biologists, we would not lose hope in saying that we are at our magnanimous best and not very far in bringing intelligent and sophisticated bioinformatics tools and use dynamic approach where *in silico* models are widely exploited. Here, acceptance and use of computer-based and *in vitro* methods in fundamental research in testing chemicals, medicines, applying biostatistics through experimental design are inevitable thus raising questions—can animal models be replaced?

To address this key question, opinions were raised through bioinformatics.org online polls and an extensive discussion was organized through SAB forum (http://www.scienceboard.org). Bioinformatics clubbed with SB have been practically two fold as the practitioners understand how molecules work *in silico*; the how of chemistry between biology and information technology and importantly see how genes or proteins could be predicted heuristically or nonheuristically algorithmic, thereby we could approach the wet-laboratory before hand in a more organized mode. Bioinformatics, by and large, has become an enforced tool in today's full-bodied molecular biology. So the popularity is not for professionals from bioinformatics only. In neutrality, not every person from bioinformatics will have these types of statistics, but let us judge ourselves closer to getting hold of computational biology or bioinformatics by following the three Rs before experimenting in the laboratory!

We leave our thoughts with the following quotes by Dr Peter Mansfield (GP, and Founder-President of 'Doctors in Britain against Animal Experiments'.) Animal Experiments in Medicine: The Case Against, May 1990:

There is no comprehensive animal model for humankind... The truth is, and always has been, that the first clinical use of new medication in human patients provides the first reliable clues as to what can be expected of it. Premarketing research on animals is a lottery; post marketing surveillance comes too late for the first human victims of side-effects.

1.11 Addendum: Results of Poll @ Bioinformatics.org

Only 22 % of the people voted for yes when asked could computer models someday replace humans in clinical trials while 50 % voted negative while 28 % have no hopes at all.

1.12 Opinion of Few Scientists on Bioinformatics Influencing Animal Experimentation

Lorikelman opined that an Artificial Intelligence—Turing test could be an option that predicts human behavior. 'Problem will be the unexpected interactions between pathways or organ systems that we might expect to see in a fairly large number of people that therefore could be observed in a clinical trial and so I don't think models will be replacing trials soon. Second big problem will be the occasional catastrophic individual reaction that some people have to a drug—difficult to model' was what Lorikelman had to say. Furthermore, he feels with no decent models available, the information about human metabolism and human immune reaction cannot be understood.

R. Wintle added saying that it could be problem with regulatory agencies buying computer models as they may not seem to work. Also, uncontrollable environmental effects on drug efficacy, and potentially also stochastic effects are further hindrances to model the animals.

Jooly opines that improved computer models may be extremely helpful in terms of 'reduction,' but feels she cannot imagine that they will ever be good enough for complete 'replacement'.

R. Stevens says, 'I've seen too many people say that we can someday understand "gene products" by just looking at the DNA sequence to fall for this idea. As soon as you think you know all the variables needed to understand something *in silico*, someone will discover that the variables were all for one gender, race, age or whatever and the whole thing will be wrong. Not those clinical trials are perfect. Even if you do everything you can to test a new drug/treatment, there could always be something out there that wasn't predicted by the trials'.

Chapter 2
Ten Reasons One Should take Bioinformatics as Career

2.1 Bioinformatics is Challenging and One is Free to Respect Open Access

Bioinformatics has enabled wet-laboratory biologists respond to the demands of ensuring quick results for the research done in the wet laboratory. Making the wet - laboratory biologists introducing some methods and predictions so as to lessen the scale of experimentation would not only help the researchers for educating undergraduate students but also allow them to move forward. As many researchers feel bioinformatics to be not a traditional bioscience, it reflects the growing modularity of biology even as it is equally diverse and have wide array of solving biology problems.

2.2 It Delves into Predictions but *bona fidelity* is the Means for Predicting Genes

Bioinformatics tools are sometimes trivial but are based on lots of predictions. For example, a protein multiple sequence alignment would delve into the status of which sequence might have evolved first. Whether or not the sequences are related can be interpreted using BLAST against a reference dataset, the annotation associated with potential matches can therefore be used to identify the gene sequences. However, the bona fidelity of the sequences will be questioned, if we do not use such sequences which can be aligned with the query protein matching the original query protein.

Fig. 2.1 A cartoon depicting the importance of low-cost stride of sequencing

2.3 Intelligent and Efficient Storage of Data is the Key

To benefit from the bioinformatical opportunities while overcoming the challenges in post-genomic era, several models are adopted which demand efficient Information Technology (IT) approaches. These are to be integrated for efficient storage and intelligent data management. Many storage approaches have been deployed widely over the past few years which are insufficient to meet emerging storage and data management challenges, the approaches that treat data in the form of virtual computing are to be discussed.

2.4 Development of Tools and Programs Making Wet-Laboratory Biologists Ease their Experiments

To convince a biologist is like winning an idea. One of the best examples that one could pertain to is designing the primers. Although there are tools available to design primers for a sequence in an efficient way, it is far less useful as they may not get desired amplicon every time. There are hardly any tools these days which present proof of concept by setting up experimental validation of functionality. As all possible primers are individually analyzed in terms of GC content, presence of GC clamp at 3'-end, the risks of primer–dimer formation, and intra-primer complementarities, a wet-laboratory perspective for designing software would make the researchers inviting to take up bioinformatics.

2.5 It is Multifaceted and Brings Networking Among Cross Disciplinarians

Multifaceted disciplinarians and scientists should join hands for a better Science. Take the example of developing a web server. The biologist would interpret the background data, a machine learner or a mathematician would think of using support vector machines. Bioinformatics is a multidisciplinary field and requires people from different working areas. It is the combination of biology and IT to discover new biological insights and there is an utmost necessity of tools that helps them to work together.

2.6 It may Partly Influence Animal Experiments

Decades ago legislation on the use of animals was enacted in many countries involving three Rs': Reduction, Refinement, and Replacement of animal models. Ever since this was enacted, there was a sudden buzz about laboratory animals and their use to be reduced, refined, and replaced wherever possible, for ethical and scientific reasons. The three Rs concept was put forward by W.M.S. Russell and R.L. Burch in 1959 in The Principles of Humane Experimental Technique. A great detail on the three Rs was reviewed by many in the interest of good and humane science. The word 'alternatives' came into use after publication of the book 'Alternatives to animal experiments' by David Smyth, in 1978.

2.7 Bioinformatics Curation, not Annotation is the Key for Databases

The knowledge base (KB) construction and semantic technologies (ST) have been intensely shown great importance in the growth of bioinformatics and computational biology. However, the KBs ensure manual curation is not sufficient for annotation of genomic databases.

2.8 Use of Bioinformatics Methods Propel Contract Research Organizations

A contract research organization (CRO) in bioinformatics is the need of the hour especially in clinical research. A CRO can provide services such as commercialization and technology licensing pharmaceutical, assay development, preclinical research, clinical research, clinical phases management, and vigilance. Many CROs specifically provide support for drugs even as software must be made available using world wide web.

2.9 Instigates Core Programmers and Developers to Enthuse Bioinformatics in them

Core programmer/developers' would not have any interest in biology unless one gets motivated from bioinformatics in application of their projects. They can only be recruited as mere developers and possibly would be very good listeners as they can easily understand biology of it. After all, a circuit diagram in computer chip set is similar to the biology system. Isn't it?

2.10 It is Dynamic and so is Inviting to be Entrepreneurial

As we have documented extensively, R and D through education would have substantial returns in two forms: privately and socially. The cross-section of researchers could fully utilize an individual's education by becoming entrepreneurs, with returns lower than in the perfect match, they are still substantial. Moreover, let there be looking back at success rate as it is seldom last for an entire working life. Educating entrepreneurs is in the high end of the interval and so, investments from angelists, brokers, and venture capitalists are important for offspring benefit, as more educated parents have more successful children.

Chapter 3
Developing Bioinformatics Skills

Cs gets degrees?

> *You never get a second chance to get a first impression ~ Oscar Wilde.*

Can a mediocre student raise high above standards in bioinformatics? Did achievers succeed abruptly in a first shot? No, they have had tasted lots of failures. That said many graduate students who have taken bioinformatics as a taught program in developing countries have had problems in identifying job prospects.

P. N. Suravajhala, *Your Passport to a Career in Bioinformatics*,
DOI: 10.1007/978-81-322-1163-1_3, © Springer India 2013

So what would be the fate of the aspirants with the third grade? In bioinformatics, we believe they stand tall. Remember as a bioinformaticist, it is your duty to remain multifaceted, think multifaceted irrespective of the stances that you take. Plan your curricula properly. Furthermore, grading standards may become even looser in the coming years, making it increasingly more difficult for graduate schools and employers to distinguish between excellent, good, and mediocre students.

The following are the two plans for the ones who do not get the job offer:

1. Plan for a management degree in bioinformatics: What is seen in a researcher is how /she manages a laboratory. Managing a biotechnology laboratory is a very important entity for a researcher and so as /she understands the intricacies of the market. For example, an intriguing question one could ask is what if not? There must always be plan B. Biotechnology or bioinformatics are just tools and one needs to be an expert in using them. If there is not a research focus, one can opt for managerial programs which can be worked on the following:

   (a) Transforming biological entities
   (b) Understand the patent regime and monopolies that lead to higher costs for drugs and treatments
   (c) Understanding Intellectual Property Right System with clear background technicalities.

What is often pointed and overlooked, however, is for those who have not at all worked hard to achieve or become successful researchers. The IPR and Management courses would typically be abundant. What do you think one can make use of that? It is the student's job to find out whether or not the skills acquired from his erstwhile education interests or provide value to his successful profession. On the flipside, core competencies in bioinformatics can be increased year over year, the average GPA at universities and colleges across the nation is on the rise.

One of the other advantages is that the students may be getting better educated in bioinformatics rather 'Exceptional Mediocrity.'

While the third-graded students would find it crucial to discuss integration of biology and information technology (IT) subjects, what about the rest? And you will succeed because you have leadership and communications skills. The ability to sell ice to an Eskimo does not necessarily require college credentials. Can a mediocre student get into bioinformatics? Yes, what all matters is to be disciplined, determined, dynamic, and diligent (The four D rule).

3.1 Be Devoted

Ever since evolutionary biology was developed by Ernst Mayr, many multifaceted and scientific works have been established which effectively brought bioinformatics, one among many regularities into the wider biology and extensive post-synthesis work in systems biology. Making a chief disciplined builder in bioinformatics proves to be an important step in drawing together multifaceted disciplinarians. The bioinformaticists have an increasing sense that a 'new' biology-related 'IT' regularities was emerging that would bring together the experimental methods of genetics and T. It was not until Paulien Hogeweg and Ben Hesper introduced the term in 1978 to refer to 'Theoretical aspects of Chemistry and Biology'.

3.2 Be Determined

In taking up bioinformatics, one has to understand that determination succeeds ant forms of academia. Understand what is that you are good at, never set aback your temper, and please be advised that you need to face challenges from time to time. The greatest challenge for one would be to liaise between the native field and how acclimatizing to bioinformatics. Take up a problem, formulate it, and thereon allow the 'D' to answer yourself.

The other Ds: Diligence and Dynamism: One needs to identify a great profession in bioinformatics with the focus, diligence, and dynamism of all subjects involved with extremely high standards. This will allow us to also achieve a very good result in acquiring synergies between the multidimensional scientists and ultimately in becoming an even stronger force in the IT and bioinformatics.

3.3 Bioinformatics and the Three Cs of Research

3.4 Consistency

The first C, one has to follow, is the 'internal consistency.' This C assesses whether or not the candidate who opted for bioinformatics has chosen the same quality, skill, or characteristic. Measured by the precision, this reliable entity often helps researchers interpret data and predict the value of scores and the limits of the relationship among bioinformatical variables. Assume a researcher designs a questionnaire to find out about bioinformatical problems with a particular focus on say, cardiovascular diseases (CVD), analyzing the internal consistency of identifying the questions that fall not just on CVDs deals with dissatisfaction and this we will liaise on the questionnaire focusing on CVDs. What would be the perchance of CVDs making the problem in the future? Are there any genes that make up the problem would mean that the researcher is trying to put up a brave front in understanding the problem better?

Continuity of Research Efforts: Tracing the communication of scientific and technical information in research is characteristic of two media—the meetings and networking and the journals/way to publishing. As research is a cyclic process, researchers are the producers of scientific information. The continuity of research belies with the sources of information used by active researchers in correlation with the current research taken by them. It is nice if the authors/researchers stick to their same area and develop their specific subareas of research while involved in publications of their article wherein continuing authors could publish a subsequent similar/different article in the same area as their original article. However, there is a lack of continuity of inquiry and progress to move forward in research as authors seldom publish in the applied research but stick to reporting the results of a single study.

Credibility: Biological credibility is what one needs to describe to move forward. Whether or not, there is a logical entity proposed in research is a system with which one can have causal effect. This proposed mechanism should be consistent with the current understanding of biology. For instance, many researchers were averse to discuss synthetic products especially coming out of artificial expression, which are proteins developed in the laboratory. Scientists rejected the hypothesis, because there was no known explanation for how proteins could copy themselves. This also invited discussions from statesmen on ethics of making proteins and their artificial expression. Furthermore, the use of prions and their spread of an infection to a new individual have changed the scope of protein studies on how they copy themselves in newly infected victims. A biologically plausible mechanism is at stake while we understand how synthetic biology takes shape. So the need of the hour is to make credible research.

3.5 Hate Wet-Laboratory Work?

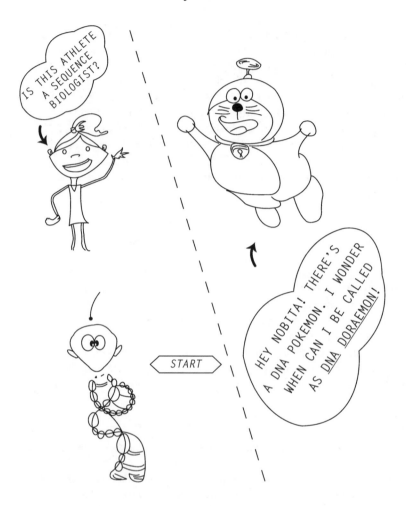

3.6 Coping the Pressure of Experimental Work

Coping the pressure of experimental work by judging your fairness for research. Ask how good are you able to cope with the loss of your results if you are not able to justify it substantially. You could also ask what type of nonsense experiments would have no results. Are experiments always predictable? Laboratory work is fairly time-consuming and labor-intensive even as it involves bizarre working hours. Although time-consuming, it would be very rewarding to work in areas of very core/integrative biology, especially looking at fundamental questions on how on earth life evolved and how bioinformatics can leverage huge data? With many researchers in the laboratories renowned for doing extra and long hours with intense work ethics, one needs to ask whether or not it really works in working for long hours, thereby doing better science. Working strenuously in the laboratory is considered one of the reasons why many people hate working. That said, coping the wet-laboratory work sometimes is seen as 'slave-drivers.' Thanks to bioinformatics, many researchers started using the tools thereby lessening the scale of experimentation. One would advocate the importance of spending time away from the laboratory works for keeping fresh while the others would claim the workaholism is the only way to succeed. Let there be hope and belief that the scientific community will continue to consider wet-laboratory scientists toward creative endeavor, and diversity in the workplace is surely to be encouraged.

The following is the classical example of how bioinformatics has lessened the scale of experimentation. The analysis of proteins in peanut leaf has been shown to have a direct approach to define function of their associated genes. Proteome analysis linked to genome sequence information was deciphered wherein two-dimensional gel electrophoresis in combination with *in silico* based sequence identification was used to determine their identity and function related to growth, development, and responses to stresses (Katam et al. 2010). Furthermore, upon verification, we transferred the protein interactors from Arabidopsis to peanut which has enormously negated the idea of running bioassays like pull-down assays to check for protein interaction partners. In this process, we could be able to identify some potential proteins including RuBisCO, glutamine synthetase, glyoxisomal malate dehydrogenase, oxygen-evolving enhancer protein, and tubulin. bioinformatical analyses have not only further allowed us to understand how these groups of proteins were accorded to their cellular compartmentalization and biological functionality, respectively, but also led to the development of protein markers for cultivar identification at seedling stage of the plant.

3.7 From 'Hands-on *in vitro*' to 'Hands-on *in Silico*'

It would be fugitive to say that all wet-laboratory work can be formulated for predictions *in silico*. It is certainty that bioinformatical predictions help the wet-laboratory biologists to reduce the time frame set for experimentation. What and how good bioinformatics can help experimental scientists to overcome the stress and cope it need to be a foregone conclusion.

Many researchers consider bioinformatics a hackneyed term and do not understand the application of it. There is a lot to read about bioinformatics, more than a tool. bioinformatics has helped wet-laboratory biologists and other cross-disciplinarians tremendously in the last decade even as there is an increased automation in the generation of data from sequenome to phenomes using genotypic information. bioinformatics has connected biological data to hypotheses by providing up-to-date descriptions in analyzing sequences. This has further allowed sequences to elucidate the literature and study evolution of organisms.

The application of high-throughput DNA sequencers has already provided an overload of sequence data from analyzing DNA and protein sequences, from motif detection to gene prediction, and annotation to curation. There has been a wide focus on gene expression analysis from the perspective of traditional microarrays by an introduction to the evolving field of phenomics. Furthermore, there are associated mining tools which are becoming increasingly essential to interpret the vast volume of published biological information, while from a developer's point of view, one needs to describe the various data and databases toward common programming languages used for bioinformatics applications.

The following are a couple of case studies that show how one can correlate *in silico* approaches to wet-laboratory. While the first explains the need of statistical inference for an experiment, the second combines the bioinformatical predictions using an already existing wet-laboratory data.

3.7.1 Correlating and Identifying Statistically Significant Causality Data

Correlation does not imply, hence causation is the paradigm behind understanding certain data, viz. biomolecules. From formulating a problem to separating correlated data, causality is a major difficulty in understanding complex difference of molecular biology from a systems level. So what if we have a data correlated to different groups showing relative antigenicity data linked to them? Could we assign likelihoods of causality to these groups? For example, an antigen A is a causal for another antigen B in a group X, the reverse causation B being the causality of A may not hold true. Here, we consider antigenic data linked to various subpopulation of HIV infected ($\pm$patients) by developing a strategy to determine causality of an antigen-specific data in various subgroups of diseased population. This white paper may be used as a measure to predict the antigens targeted to various groups, here more precisely 'causality.'

Problem
Identifying variations in the antigens susceptible to various diseased group of subpopulation provides functional information on how genes lead to disease. Let us consider the below-mentioned data which consist of the peripheral blood mononuclear cells (PBMC) from three groups which were stimulated with different bacterial antigens. In turn, the cytokines produced by the PBMCs play an important role in the immune system. Based on the production of cytokines, immunogenicity in these groups could be better understood. We propose a method to sensitize and identifying positive predictive accuracy (PPA) using *t* tests in predicting the antigens (causal data) immunogenizing the various groups.

3.7.2 Brief Methods

Dataset

The three various groups of HIV that were analyzed are as follows:

1. HIV positive
2. HIV Control
3. HIV negative and other 'diseases' negative

Statistical Analyses

1. Positive Predictive Accuracy:

The sensitivity and specificity do not always provide the probability of correct hypothesis. At times, we must approach the data using predictive accuracy. The PPA is based on the probability of the antigen to be more immunogenic against a certain group. We used the following three calculations to perform the prediction:

(a) Sensitivity is the proportion of true positives correctly identified in the data
(b) Prevalence or likelihood is the number of samples/total number of samples within the group
(c) Positive prediction accuracy or PPA is calculated using the formula.

$$\text{PPA} = \frac{\text{sensitivity} \times \text{prevalence}}{\text{sensitivity} \times \text{prevalence} + (1 - \text{specificity}) \times (1 - \text{prevalence})}$$

However, since we have taken the proportion of true positives to be maximum while ignoring the true negatives, specificity remains out of question, thus making

$$\text{PPA} = \frac{\text{sensitivity} \times \text{prevalence}}{\text{sensitivity} \times \text{prevalence} \times (1 - \text{prevalence})}$$

$$\text{PPA} = 1 - \text{prevalence}$$

2. The *t* test of significance: Student's *t* test was used based on sampling the three groups as they have unequal variances. In particular, this test is sensitive and could be used to yield a better probability, a value better than PPA which might provide ample evidence to support the above hypothesis. While this test is the standard test for calculating the relative efficiency of other tests (in this case PPA), it also requires the most stringent assumptions.
3. Friedman test: The Friedman test is a two-way analysis based on ranks which models the ratings of a (rows) 'antigens' on b (columns) 'Groups.' The test parameter here *W* is called Kendall's coefficient of concordance.

W = Sum(Rct^2) * 12/(a^2 * b * (a^2 − 1))−3 * (b + 1)/(b − 1) //
Kendall's coefficient

Q = a * (b − 1) * W // Chi square

Degrees of Freedom (DoF) = (b − 1) // Number of columns − 1

Furthermore, the mean Spearman rank correlation coefficient (Rsm) between all the rows could also be identified using (a * W − 1)/(a − 1).

3.7.3 Interpreting the Results Based on Preliminary Analyses Using PPA

In the data provided, we have used all the three groups and scored the sensitivity, prevalence, and the PPA. Interpreting the probability using prevalence is crucial and was carried out using the total number of subpopulation in the groups. If the prevalence is low, the PPA is high, which clearly indicates that the data constitute all true positives. Had specificity been inclusive, it is highly inevitable that the results will be false positives. However, a high PPA may indicate that it is statistically relevant to find antigens specific for immunogenicity, but it does not necessarily indicate the presence of immune response. Further analyses on detection and elicitation of immune response using cytokines were carried out wherein the preliminary analyses using t test of significance showed that the groups 1 and 2 are statistically more significantly compared to the third group. Considering the fact that the group 2 is a control dataset, we would, however, find it to be significant; hence, a test of hypothesis was carried out to find statistical significance.

T Test of Significance

- For groups 1 and 2: t = − 12.01, DF = 25, p <= 1.615e-09
- For groups 2 and 3: t = 12.59, DF = 25, p <= 1.045e-09
- For groups 1 and 3: t = − 13.58, DF = 25, p <= 5.352e-10.

Where N is the total numbers (population), DF is degree of freedom (N − 1), and t, the test parameter = (Mean/SD) * sqrt(N) and p, the probability. Since t is substantially same for groups '1 and 2' and '2 and 3,' the group 3 is ignored for a very less probability.

Friedman Test

Kendall's coefficient W = − 3.29

   Chi-squared Q = −164.44 ∼ X^2

   Degree of freedom DoF = 2, p <= 1.00000

The level of significance, p <= 1.00000, given above is based on an approximation of the chi-squared distribution. Another statistical significant concordance is if the test parameter Q is high (i.e., statistically significant), then the columns are known to be different and the rows are correlated which in our case, the dataset holds true (Q being relatively high).

3.7.4 Predicting the Antigens Immunizing the Groups

From the above analyses, it is clear that the three methods we employed are independent with each other and are sensitive to apply statistical significance. Further, sensitivity in groups 1 and 2 when averaged (0.25 and 0.58, respectively) (see data below), it sets the mark for identifying the candidate antigens. Hence, the antigens whose sensitive value are par below the above-mentioned respective values for the groups are discarded (see the other data tabled) while the rest are used to identify cytokines against the antigens that play an important role in the immune system.

3.7.5 Conclusions

The antigens, viz. Z, A', and B' are specific to the two groups 1 and 2. Based on the values obtained, the production of cytokines specific to the group as against the antigens can be identified from the data. Furthermore, these could also be used to integrate co-expression networks and genotypic data. If the data constitutes expression traits, we could establish statistical significance using the methods discussed as above. The causal predictions that were made (see Table 1.2) could in turn be used wherein the data may be divided into training and testing data randomly in 8:2 ratios. To avoid the selection bias, training set cross-validation of 10-fold could be carried out producing an accuracy. While testing on remaining 20 % test dataset, the predictive accuracy could again be established using radial basis function support vector machine (RBF-SVM) kernel. The use of SVM-based classifier gives the best result among all other classifiers, but the limited accuracy performance might challenge the machine learning classifier.

3.7.6 Bottom Line

1. **Which methods are readily implemented and able to extract biologically relevant causal connections among genes?**

   - PPA, *t* test, and Friedman tests
   - Support vector machines (SVM).

2. **If a method employs data normalization, what are the strengths and weaknesses of the normalization algorithm in terms of facilitating data analysis and interpreting the data and results?**

   - Strengths: Sensitive and highly accurate, and easy to store data
   - Weakness: Loss of data as it is pretty difficult to ascertain.

3. **What are the pros and cons of each method?**

   - PPA

Pros: Direct disease could be ascertained
Cons: It is extrinsic, meaning it is always depended on other factors, viz. prevalence.

   - The t test

Pros: Sampling huge datasets, correlation
Cons: The user must be aware of how big the sample is and what for the data are to be used.

   - Friedman test

Pros: Easy ranking and nonparametric test
Cons: High rate of false positives and dependent on other datasets

   - SVM

Pros: Prediction accuracy is always high
Cons: Cannot be used for training if the data are less and mediocre.

4. **Are there pitfalls to avoid with a given method, or circumstances under which the method may be less reliable?**

   - Using SVM, the number and inappropriate set of descriptors and multiple target classes make this method more cumbersome to act as an efficient tool especially to predict genes

5. **What criteria were used to evaluate and rank the methods?**

   - Ranking based on the existence of antigens specific to a group and whether or not a cytokine particular to the group is produced.

3.8 Case Study on Nematome: Protein Interactions Specific to Parasitism in Nematodes

The nematodes, commonly called the roundworms (belong to phylum Nematoda), are the most diverse of all animals. Of over 28,000 described so far, an approximate 16,000 are parasitic. The parasitic nematodes especially those that from plants have not yet been known better. Furthermore, genome sequences of the plant-parasitic nematodes are just beginning to yield results even as RNAseq (transcriptomic) analysis is being done by us and many other laboratories. With *Caenorhabditis elegans'* genome completely sequenced way back in 1998, nematode genome sequences hold a great promise to understand the umpteen nematodes' genomes waiting to be sequenced. It has been known that the genomic repertoire and gene-centered density is roughly about 1 gene/5 kb with 24 % introns on an average across all nematodes. While many genes arranged in polycistronic series of operon model, there holds a greater importance to understand mitochondrial genome as well owing to identification of parasitic genes in nematodes. The RNASeq and RNAi studies have started yielding results too with biology curators appraising the set of known genes even as the predictions need to reach consensus along with the flourished datasets of ESTs, RNAseq, and genomic repertoire. So that begs a question whether or not any commonalities of all these genes are known across all nematodes? The answer belies in how and what kind of organisms are these: parasitic, nonparasitic, entomopathogenic parasitoids, non-entomo nematodes, etc.

For example, *Caenorhabditis briggsae* genome and further comparative genomic analyses determined the novel gene sequences from the same genus nematodes such as *Caenorhabditis remanei, Caenorhabditis japonica* which further enthused knowledge that they might be less likely to complete and remain accurate than that of *C. elegans*. That said, the worm-based database is void of many plant-parasitic nematodes. Further understanding to nematodes has revealed that there has been an accelerated rate of evolution in the parasitic lineages while several phylogenetically ancient (Read parasitic) genes might have been lost and found elsewhere across all nematode species. In that process, RNA interference (RNAi) experiments leading to gene loss of function were done even as researchers were able to knock down about 86 % of the $\sim$20,000 genes in the worm with an established functional role mounting to 9 % of the nematode genomes on an average. However, the story does not end here with ascribing function to the genes as the aforementioned methods are mediocre and involve lots of false-positive datasets. With systems biology burgeoning, there is a need to understand the how of interactions in nematodes. The *C elegans* protein interaction network (PIN) was a masterpiece of genomic catalog of protein–protein interactions (PPI). The interactions have been established based on the small-scale experiments while there is need to complement the *bona fide* interaction studies with large-scale datasets. Predicting the PIN across individual nematode genomes involves lots of experiments, reactions, and importantly wastage of man-hours. Therefore, we wish to propose a uni-comparative biology approach to predict PPI across ergonomically important nematodes of parasitoids, viz. root knot, migratory, and most damaging. Nevertheless, the interactions can be ascertained and the function can be better ascribed across these nematodes wherein we would identify the proteins involved in parasitism. That said, we propose a word called nematome for those set of (most commonly occurring) proteins implicated in parasitism.

3.8.1 Methods

Wet-laboratory experiments

Plasmid Construction

(i) **Two-hybrid constructs**: The cDNAs encoding full-length parts will be amplified by PCR using gene-specific primers containing REs with EcoRI and BamHI restriction sites. The PCR products would then be digested with REs and then ligated.
(ii) **Plant/nematode expression constructs**: The constructs used to transiently express the interactor proteins will be based on plasmid, viz. pMON999, which will contain the proteins specific to the promotor and terminators (van Bokhoven et al. 1993). The cDNAs of the positive clones will then be excised from the vectors using EcoRI and ligated in the EcoRI site to allow expression of the interactors tagged with other proteins.

3.8.2 Interaction Analyses

- **Yeast two-hybrid screening (or co-immuno precipitation):** The initial experiments will be enthused with a *C. elegans* cDNA library in the desired vector (Promega/Clontech). This library thus should be constructed with all possible independent cDNAs. Colonies are then selected on agar plates lacking histidine, tryptophan, and leucine over a 7-day period while positive yeast transformants will be picked up and replated Gal assay. A positive interaction can then be determined by the appearance of blue colonies and the plasmids can be isolated. In subsequent experiments, bait and prey constructs containing full-length proteins or domains will be analyzed and later can be transfected.
- **Immunofluorescence labeling and fluorescent microscopy:** This can be performed essentially based on the aforementioned results and these can be imaged using fluorescent microscopy.

3.8.3 Dry-Laboratory/Bioinformatics

1. Traditional mapping of interactions for parasitic genes with respect to functional annotation.
2. Interolog mapping: While it is known that the orthologous genes are highly conserved between closely related species, we presume that the systems might utilize the same genes and share interactant information across the orthology

datasets across different organisms. However, it does not necessarily mean that the amount of sequence conservation is directly proportional to interaction even as certain studies comparing high-throughput data including expression, protein–protein, protein–DNA, and genetic interactions between close species show conservation at a much lower rate than expected. We would like to identify those parasitic genes from step (i) and show that conservation is maintained between species albeit through network modules. Further, we would like to employ confidence score for interactions based on available experimental evidence and conservation across species. (Please refer flow chart below.)

3. Filtering the datasets and reaching the consensus: We would then filter interaction datasets and integrated them with a high-confidence interval thereby reaching consensus. This would ensure that the estimated size of parasitic interactome of nematodes (nematome) would have approximate number of interactions. Comparison with other types of functional genomic data would show the complementarities of distinct experimental approaches in predicting different functional relationships between genes or proteins. Finally, we would like to compare them against different tissue-related proteins with respect to co-immunoprecipitation (CoIP) assays. A further re-examination of the connectivity of essential genes in nematodes could support the presumption that the number of interaction partners can accurately predict whether a gene is essential and if essential to which organelle. This would yield organelle proteomic analyses. In conclusion, our analysis should facilitate an integrative systems biology approach to elucidating the nematode cellular networks that contribute to diseases (Fig. 3.1).

3.9 Tips and Traps in Writing a Research Article in Bioinformatics

Many consider writing the article/proposal to be the toughest and perhaps most boring part of the entire report-writing process. The best way to start a project work especially PhD is to start a review. The student must have a firm grasp on the topic that they have been introduced to or the work they have been acquainted, and therefore put in countless man-hours of the literature review, formulate a problem, and finally then submit the results in writing which is a difficult task, to say the least. The following guiding principles could provide the reader on how of taking up the problem, defining it, and the logistics of the write-up, etc.

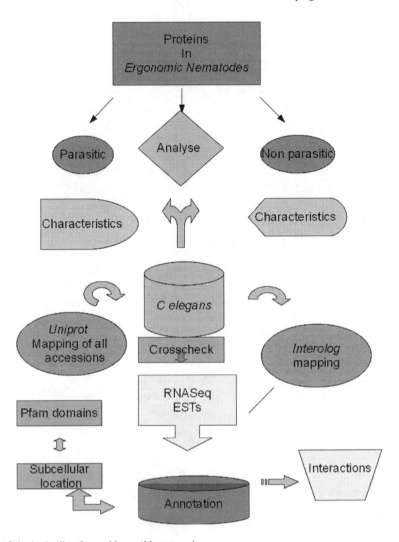

Fig. 3.1 A pipeline for making a 'Nematome'

3.10 Convert Ideas and Thoughts into Action for a Strong Problem Formulation

There are changeable ideas always about how long or how short a review/paper proposal should be. Conversely, the proposal can be called as white paper if it is addressed during its preparation. But please ensure that in the former, all the points and ideation process are covered while in the latter, a concrete dialog of proposal should be summarized.

3.10.1 Address the Problem Well with Subheadings

A. **Background**

Background is an important summary of the major points addressed by the erstwhile researchers. These questions behind your research could be good compendium to address the current problem you are to address in the future and provides the context of those questions within a larger academic framework. This precisely is a kind of pre-introduction and one should be able to see who read the introduction should be able to understand what you are attempting to discern through your research and writing.

B. **Introduction and Review of Literature**

This portion should address the scope of research while listing major findings. Whether or not one sets out a specific portion of the proposal for a literature review is unto the discretion, it would be nice if a fair problem is specifically indicated wherein bioinformatics has been employed to lessen the scale of experimentation. Some points on introducing a pictorial representation or Gantt chart would also be inviting wherein some of the results can interlace within the other major portions of the intending proposal. Regardless of how one would present the literature review, we could describe the findings of the review specific to the importance of the problem chosen in the area of custom research.

C. **Problem formulation and Objectives of the study**

It would be nice if the researcher describes how bioinformatics can leverage issues as depth and provide the background and particular context of the problem in relation to the particular academic field. The objectives can be described on a point-to-point basis not making up to one page.

D. **Materials and Methods**

What is the overall plan of the experiments that will be done and why planning these bioinformatical predictions and experiments' including annotation and curation is an important element of the dissertation. All the detailed methods of research to be demonstrated relating to the question and problem formulation should also be addressed.

E. **Results and Discussion**

This should specifically focus on what was aimed during the dissertation frame, and the results, although not preferentially discussed as per the objectives, may include the contents with pictorial and graphical representations with concise statements debating and reaching the consensus of the objectives of the study. It would be nice if the points are split and an appealing statement can be made to be inviting for the reader to ensure there is a flow of very good reading. Many a time, the reader falls in trouble in not understanding the scope and depth of the problem. Especially, if it were bioinformatical predictions, what makes the predictions *bona fide* would mean that the works are more precisely specific and with respect to the growth of the works.

F. **Conclusions and Future Directions**

Conclusions typically are written shorter and play an important role toward bringing together the main areas covered in the erstwhile analyses as described until the Results and Discussion. Furthermore, it would also work on giving a kind of pre-final comment or judgment making suggestions for improvement and speculating on future directions. Although conclusions are likely to look more complex, it is to be noted that the significance of the findings and recommendations for future work are to be brought to the notice of the reader in this section wherein important implications are to be covered. The future scope and the upcoming challenges negating the line of false positives would really provide the reader and the dissertation very inviting not only toward the research but also for the reader who can cite your work in many a way.

G. **References/Bibliography**

Making a list of all source materials and properly formatting them in whatever academic style is required for a complete dissertation which will address the problem of writing with elegance.

3.10.2 Plan your Next Steps and Always give Plenty of Time

One good thing always takes us move forward is to plan the way we write the report. As bioinformatics involve lots of predictions, the wise thing would be estimate how many hours/days you can start working in writing the manuscript. The last thing at the last minute is to understand the fact that the write-up or proposal even takes up to 6–12 months to complete or sometimes even longer. Always ensure there is room for tackling questionnaire, discussion with your peers, and importantly call for constructive criticism.

3.10.3 Discuss with your Peers

Always broader understanding of research topic helps because when it comes to dissertation proposal, many a proposal is negated and they eventually get off the ground even without an adequate review. That will allow us to invite contacts in the form of peer reviewers and committee members to prepare the report which plays a vital step to stay in touch for the future. Furthermore, they are available anytime for open advice anytime. There is nothing wrong in asking copies of previous reports, so that these which were once approved would allow you to prepare with ease and satisfaction.

3.10.4 Accept Constructive Criticism

Peer reviewing and the purpose of a support group are to ensure we read each other's work and give feedback. In doing this, we will not only help improve each other's writing but also allow us time to read while evaluating the work and later providing the feedback. That said, one need not be an expert and so let not content or vocabulary intimidated. Even if not familiar, a strong argument can be made on how the manuscript has been written and addressed within the scope of the journal or subject, etc.

3.10.5 Publish or Perish is the Key While Citing and Cross-Referring Other Articles of Interest

After research is well taken into, it is recognized only on account of publications. The world knows only after the work is published where people come to know about it. Adding references to such work into their publication text and list of references can be seen as a kind of good normative citation. That said, references can be divided into single units, whereby each reference turns into a citation that can be aggregated in many different ways, forming a wide range of citation impact factors/indicators (CIF). It is like many such articles are references and indexed in citation indexes, such as the Thomson Reuters' Web of Science database even as many online repositories such as Google Scholar, BioMed Experts in the form of 'scientometrics' work for the cause. The key here is in the competitive field, publish, or perish.

3.10.6 Peer Review Holds an Important Community Service

Peer review is the key! The more you review, in all likelihood, the more you will be asked to review. Often times you may be asked to review boring papers that are of no interest to you. While it is important to serve as a reviewer, only accept papers in which you are keenly interested, because either they are close to your area of research or you feel you can learn something. You might say that should I not know the work very well to be a reviewer? Often a perspective from someone in a slightly different area can be very effective in improving a paper. Editors would of course like to see your review papers even if you are not particularly interested in them, but the reality is that good reviewers must use their reviewing time wisely.

Chapter 4
The Esoteric of Bioinformatics

There are few people who can understand the intricacies of bioinformatics. Through myriad bioinformatical predictions, the wet-laboratory observations, and experiments can be developed to illustrate ideas on the problem formulation based on genes or proteins. Topics include the biology of system, sources, and effects of bioinformatical predictions, characterizing the uncharacterized genes and genomes, and applications in medicine and agriculture.

Basic skills in integrated biology would be a plus which include knowledge of biochemistry, bioinformatics, and molecular biology. One need not have to be an expert in all, but it helps to have at least some background in each field. Amazingly, bioinformaticists hail from many other disciplines such as Statistics, Computer Science, and Mathematics. The smartness in getting into bioinformatics depends on how good the researcher is getting acclimatized to through experience he/she had gained in his/her field. We will provide tips on how to be a winner in bioinformatics using bioinformaTICKS (*Please see Frequently Asked Questions*). Through limitless observations and discussions on experiments, we develop and illustrate the following ideas on how esoterically bioinformatics can be in nutshell.

4.1 Bioinformatics Market: Hype or Hope?

The last couple of decades have seen bioinformatics market significantly evolving across the world on back of rising omics industry. With increase in application of these omics-es in biotechnology, there has been a commercial market for bioinformatics worldwide. With declining costs of per-base genome sequencing, introduction of next-generation sequencing (NGS), widespread interest in Genealogy, micro-RNA research, introduction of aptamers replacing antibodies, etc. public and private sector investment has given a significant boost to the industry. It has been known that the biggest field of the global bioinformatics industry has been into sequencing, software, and services with respect to IT infrastructure. With the software segment improving its share, database market will suffer the downturn due to the increasing popularity of innovative analysis software including that of

P. N. Suravajhala, *Your Passport to a Career in Bioinformatics*,
DOI: 10.1007/978-81-322-1163-1_4, © Springer India 2013

SAS/inbuilt software for companies. Applications of bioinformatics in genomics, proteomics, and pharmacogenomics have furthered genome studies which completely transformed basic research. For example, diagnostics specific to cancer has become the leading therapeutic area wherein bioinformatics is a big support, be it outsourcing or finding drivers pushing the market. As the market has been witnessing the launches of key bioinformatics products and services in various areas, these developments might impact the biotechnology industry's future performance and therefore, competitive landscape for this need to be done.

4.2 Decoding Genes Using Genealogy: What Bioinformatics Can Do?

'DNA transcripts RNA and RNA translates proteins' have been the central dogma of molecular biology and with the identification of genes contributing to diseases well represented through central dogma of bioinformatics using *sequence predicts structure predicts function*, direct method for discovering the molecular pathways involved in their pathogenesis or function of these proteins *per se* could be interesting to delve. What interests researchers to delve into the genome? Let us imagine 10 years down the line what are the diseases one could predict before hand and perhaps have a diagnostic moiety represented so that humans can beware of? That said Iceland based decode Genetics have virtually established this using a built-in system of information linking medical information from patients. However, there are lots of ethics that one needs to exploit and understand and use this interpretation with a specific informed consent even as disease-by-disease studies and molecular genetic information could be ascertained.

4.3 Communication Between Organelles and the Genes

Communication between organelles and the genes can be best referred with respect to systems biology context. A system can be better described as entity constituting major components and minor components. While major components, viz. tissues, organs, and organelles, compete for space, minor components such as genes, proteins, and enzymes compete for interacting with each other or compete with the analogous components. This competence results in ascribing function. One of the major competitors is the presence of subcellular sorting signals for the proteins localized to different organelles. For example, researchers have identified key protein determinants targeting mitochondria. These protein determinants fall into at least four subclasses of mitochondrial-targeted proteins containing targeting signals directed to different sites within the mitochondrion (outer membrane, inter membrane space, inner membrane, and matrix) by diverse mechanisms (Bolender et al. 2008). Conversely, some of these proteins are not known and are yet to be discovered, remaining as 'hypothetical.' Recently, dog was used as cancer model (Khanna et al. 2006), which enabled the researchers to understand whether there are any unknown genes or genes encoding some HPs involved in diseases (Agrawal and Khan 2007; Langheier and Snyderman 2004; Nevins et al. 2003; Workman 2003; Zoon 2004; Attwood and Miller 2002; Bottinger 2004). The dog genome study has initiated exploring some diseases whose genetic linkage is not yet known. We think that some of the orthologous regions of these HPs probably were mapped, providing clues to study important diseases.

4.4 Pull-down Assays and the Role of Bioinformatics

Earlier, researchers' idea that electrospray could spray and ionize molecules using mass spectrometry was well conceived, and the substances were analyzed to ionize them. However, subsequently small molecules would play an important role in analyses wherein samples from the patients could be directly analyzed. Researchers now have been working on similar idea with proteins. However, there are several limiting factors in measuring the mass of proteins and so is the case with sequencing peptides. The Edman degradation test has been in use to determine what amino acid it was. With mass spectrometry techniques burgeoning, tests such as ED were shelved out bringing the development of more robust technologies such as tandem MS and TOF. Thanks to bioinformatics, even there are efficient bioinformatical databases employed with MS such as Mascot, a powerful search engine to identify proteins from primary sequence databases. The idea of obtaining proteins from the gel and still analyzing them very sensitively using mass spectrometry was a part of pioneering studies by Mathias Mann's research group was back in 1996. With the proteins very difficult to get out of the gel, it has been eventually discovered that the short part of the protein sequences, aka peptide sequence tag, can be searched using bioinformatical algorithms that are well known and searched for similar sets of sequences with equivocal function be ascertained. So, if one would ask whether proteomicists' view of mapping all proteins on this 2D gel electrophoresis was deployable, it must be noted that it has its own merits and demerits. Furthermore, out of the limitations and of less potential use, variants of MS play an important role. The electron spin ionization MS has enabled characterizing the protein complexes where one can look at a small number of proteins that had some functional context; nevertheless, we are now able to do it at one go. Consider an example MLH1 which is involved in DNA mismatch repair. However, it has not been known that DNA mismatch repair exists in mitochondria, but we knew that from bioinformatical point of view and interolog mapping, we could get interesting candidates and so believe that MLH1 might be inherent to mitochondria. To run pull-down assays, we need antibodies raised against the bait (MLH1 here). The idea is that the proteins when pulled down would interact with its prey proteins and therefore, would be associated with each other and so the interactions and functions could be transferred. If you had the antibody, you could pull out not just the receptor, but the other players in the known mitochondrial pathway. Very recently, we could use aptamers which completely are sensitive, and bona fidelity is at the means of cost-effectiveness when compared to antibodies. Aptamers, until recently, were used for RNA chimeras, and not long ago we discover that these can be used for pull-down assays, especially for CoIP experiments sensitizing the tagged assays in determining the function of hypothetical proteins.

That said, determining the HPs targeted to mitochondria specific to MLH1 could be interesting. There are currently 1,185 HPs in human. While searching for human mitochondrial proteins, we augmented the fidelity of *in silico* selection strategy in searching for candidate HPs targeted to mitochondria. In this process, the human mismatch repair protein hMLH1 which we explored revealed that it is not localized to mitochondria, while the hMLH1 is known in nuclear extracts of human cells. A greater amount of gene diversity remains to be studied across many DNA mismatch repair proteins including that of hMLH1. In that process, we believe many HPs targeted to mitochondria (read putative DNA mismatch repair) could be interesting candidates to study the dynamic behavior of genes and establish how HPs are distributed and altered. The work could be interceded based on wet-laboratory and *in silico* experiments. First to check for candidate HPs targeted to mitochondria, we can run pull-down assays. Whereas, the above wet-laboratory experiments can be carried out, we can also work based on the interaction studies. We could deduce putative protein interactions that we are already establishing specific to hMLH1 and diseased candidates. From the interaction map, we could find the nearest interacting partners of the protein and then model the genes involved in diverse functions and specifically that of the HPs targeted to mitochondria. Above all, while hMLH1 is just considered as an example, we can consider any genes or set of genes to study how gene functions are altered. Finally, a web server (with script complemented) is to be developed to find how genetic variation is ascribed to genes.

As MS data analysis is endless and limitless, we feel it is quantitative especially when bioinformatics has made leap and bounds in the recent past. Previously, one used mass spectrometry, as we are trying to do on MLH1, to find and sequence a single protein. Now thereon, quantitative measurements on it can be done with the kinds of things with proteins that people have only been doing so far with mRNA and microarrays. The advantage of doing them at a protein level is that proteins are the functional agents. When you look at mRNA on chips, you have a question as to whether the change you are seeing is at the mRNA level or down at a deeper level of regulation. Maybe, it is at the protein level. Nowadays, we can read out the proteome in a quantitative way, in a large-scale fashion. Another area of work that is more specific to mass spectrometry is to look at modifications. We not only want to know what proteins are present in a sample, but also how they are modified. Are they in an active status? Are they phosphorylated, for example? This can now be done in a large-scale way by mass spectrometry. We can now look at the proteome quantitatively and examine how mitochondrial pathways change by looking at how the proteins are modified, whether they are phosphorylated. By doing so, we have a very good handle on how cells process information. This will be a big theme and esoteric in the fields of bioinformatics and systems. There is another new field called 'interaction proteomics.' Here you use mass spectrometry and proteomics to see which proteins talk to which other proteins. Could the MLH1 be used as a potential biomarker for diseases? And that is also finally coming within reach. We hope that with further development we can look at the proteins in a urine sample, for instance, and then use them to classify patients. What diseases would we be covering and following? What drugs do they respond? It is limitless, isn't it?

4.5 Say 'Ome' Using Essential Bioinformatical Indicators

Access to huge bioinformatics data is essential for understanding what kind of data would be useful to experiment in the laboratory. Thanks to many omics-es which have steadfastly been available. Among several impediments to generating data and accessing in the laboratory, an important incentive for scientists to publish research articles in bioinformatics with explicit recognition from wet-laboratory perspective, collaborate with scientists, has tacit conversations through academic communities. Ome is many or monies. Saying 'ome' is the key to become a successful bioinformaticist.

4.6 Ten Career Options to Opt Through Bioinformatics

1. Scientist

WHAT: You chose to become a Scientist.
WHO: A PhD in Biology/Biotechnologies/Informatics is needed. However, MSc's with research with exceptional track record can apply for positions too.
WHEN: A PhD with couple of years of postdoctoral experience can start applying for positions.
WHERE: In Research Organizations, Collaborative institutes, etc.
HOW: You will be responsible for the following:

   (a) Envision and build databases/web servers for wet-laboratory researchers to share their biological data. In case of hospitals, help create personalized dataware house medicines with individuals' genetic code and biochemistry. Create computer tools to track and analyze the patterns of viral outbreaks, such as flu, around the country.

2. Lab Manager
3. Professor

4. Research Fellow
5. Entrepreneur
6. Analyst
7. Consultant
8. Technology Licensing Officer
9. Business Manager
10. Research Associate

Chapter 5
Common Minimum Standards: A Syllabus for Bioinformatics Practitioners

Disseminating key survival messages on bioinformatics with emphasis on common minimum standards for bioinformatics education could bring a rise in awareness of the need for non-formal and formal education programs. The following are the brief subheads of what an undergraduate in bioinformatics could be taught:

Introduction:

- What and the how of bioinformatics.

Bioinformatics to Computational Biology:

- From a mere 'Tool' to a science.

Need for bioinformatics today.
Applications of bioinformatics in various disciplines.
Current prospects and future challenges.
Homology and similarity searches using bioinformatics.
Advanced similarity searching on the Web.
Using Blast on the Web.
Searching sequence databases to predicting structures.
Phylogenetic and multiple alignment tools:

- CLUSTAL and PHYLIP.

Sequence-based taxonomy:

- From multiple alignment to phylogeny.

Validating consensus sequences.
Primer designing *in silico*:

- Properties of a bona fide primer.

Primer designing tools:

- Primer Blast and e-PCR.

A pilot experiment for designing and synthesizing a primer *in vitro*.
Hands-on with emphasis on the participants' favorite genes.

P. N. Suravajhala, *Your Passport to a Career in Bioinformatics*,
DOI: 10.1007/978-81-322-1163-1_5, © Springer India 2013

Introduction to computational evolutionary biology.
Bioinformatics for evolution:

- Validating novel proteins.

Ks/Ka score detects evolution.
Detecting selection in sequences.
Functional genomics:

- Annotating genomes to proteomes.

Why need curation?
Validating complexity in sequences/conserved regions.
Genome projects:

- Comparative analyses.

Advances:

- The HapMap project.

Poor man's genomes:

- An EST perspective.

Expression data in genomes:

- SAGE.

Web-based practices and analyzing assembled genomes.
Systems to Synthetic Biology.
What is Systems Biology? Systems Biology of aegis.
Top-down and bottom-up.
Protein–protein interactions (PPI).
Wet-laboratory methods employed for analyzing PPI.
Interpreting the Data:

- *In silico* tools and visualizers for Systems Biology.

Introduction to next generation sequencing:

- Challenges.

Semantic technologies for biologists.
Interpreting genes and proteins using rearrangement.
Conserved DNA sequences:

- Understanding promoters and restriction sites.

Domains, motifs, patterns, and SNPs.
Comparative genomics of regulatory regions.
Introduction to Bio programming.
Subcellular localization studies and role of genes as probes.
Miscellaneous topics of interest.

Chapter 6
Colloquial Group Discussion on Bioinformatics: Grand Challenges

This is the outcome of the GD addressed by multi-faceted disciplinarians. The following text is colloquial and the reader may consider it to be the voice of all the participants.

Bioinformatics was highly evolved in early 2000 and all of a sudden fallen and not much talked by the end of the decade. Various schools and universities have recently started a high-end program on bioinformatics in Western countries where as in developing countries, the taught programs are weakened on the premise there is a demand for expeditious faculty. With researchers scaling the ladder of bioinformatical progress by leaps and bounds, there is a need to identify the why and the how of lacuna for bioinformatics. Some of the excerpts from the consensus points of the GD titled 'bioinformatics: Visions and Challenges for the next decade' are as follows.

We considered how bioinformatics may evolve in the future and what challenges and research is needed to realize this evolution. With increase of diverse resources, we suggest bioinformatics will evolve by bringing biologists together to understand what precisely the 'B' word is. In other words, we all agreed that multifaceted disciplinarians play an important role to evolve bioinformatics research in developing countries for that matter anywhere. While many researchers consider bioinformatics a threadbare term, few do not understand the application of it leaving lots to read about bioinformatics, more than a tool. How bioinformatics will help wet-laboratory biologists and other cross-disciplinary scientists was discussed in great detail. It was felt that the biologists can easily understand and interpret the results of bioinformatics compared to bioinformaticists because of the girth of understandability they have. Basic skills in integrated biology would be a plus which include knowledge of biochemistry, bioinformatics and molecular biology. One need not have to be an expert in all, but it helps to have at least some background in each field. Amazingly bioinformaticists hail from many other disciplines such as Statistics, Computer Science, and Mathematics. The smartness in getting into bioinformatics depends on how good the researcher is getting acclimatized to, through experience he/she had gained in his/her field. That said, we also discussed on how to be a winner in bioinformatics and perhaps not make paraphernalia.

P. N. Suravajhala, *Your Passport to a Career in Bioinformatics*,
DOI: 10.1007/978-81-322-1163-1_6, © Springer India 2013

While all of us agreed to the fact that the bioinformatics in India and developing countries has been hyped a lot, an Information Technology (IT) aspect of it was considered to be one of the reasons even as high salary market for bioinformatics was assumed to play wet blanket. Many people also agreed to the fact that it is not really a problem to digest the big B word, with no compulsion set to it. We also agreed that funding, reaching consensus; flexibility and collaboration are the key for bioinformatics in Indian research to move forward. On a note on whether or not we are good at writing research grants, there was a split in the opinion, wherein many opined that yes, it could be; while bureaucracy hinders the innovative and ideas. That said, we also herald the discussion for respecting open access, which is stellar for bioinformatics development.

Bioinformatics as a subject should be taught at school and college level along with major subjects or it should be a part of computer application. That said, students who come out of college or universities will have an idea about what actual bioinformatical applications can be leveraged in solving major biological problems. The government could also take initiative step to encourage bioinformatics by giving funding to the projects, facilities and positions in under graduate schools and colleges. Likewise, all faculties from could visit colleges and schools once in a month and cater to the understanding of bioinformatics in the schools. With the world facing a lot of problems in the environmental and medical issues, we reached consensus that bioinformatics is the solution through the combined efforts. Traditionally, we have a lot of acceptance to the novelty and inventory through the three C's in practice: creativity, credibility, and continuity. That said, we can grow high lest we realize that the growth of bioinformatics in India is in our hands for that maxim we need to follow: To be enthused is to be infused with life!

6.1 Opinion of Bioinformatics Practitioners

Madhan Mohan opines…

1. How should bioinformaticist make up the blend of being a biologist as well as a programmer?

Right from the career point, the candidate has chosen, the student should be preparing himself for problems specific to both Biologists and Informaticists. I suggest students give emphasis in understanding the programming logic, make an algorithm on white paper and then try compiling the program, testing it several times. Initially, there could be acclimatizing problems for students where biologists may not find programming logic interesting as most of them hail from pure biology background and not taught Maths during their intermediate schooling whereas in Indian scenario at least it has not been difficult for programmers to understand biology. Interestingly, the latter genres of students have become the most fitting entrepreneurs in bioinformatics. In conclusion, try achieving it, not thinking of failures.

2. **How best to formulate a problem for research in bioinformatics? What are the challenges one should pose?**

As a bioinformatician, please understand that there are genres of multifaceted disciplinarians who have background from Physics, Statistics, Informatics, Biochemistry, etc. Try making a preproposal and collaborate with biologists and biochemists who do not have a background in programming. That said, problems described through programming should be understood with basic logic and then algorithms could be written. The main challenge is to understand the problem, and solve it whereas if you are a programmer, simply solve it.

3. **Given the high-dimensional genomic data generated from time to time, how bioinformaticist should keep himself updated?**

Just follow the new publications that are available in reputed journals such as Nature, Science, Public Library of Science, etc. You will find the greatest updates at the three genebanks available worldwide, viz. NCBI, DDBJ, EMBL-EBI, and from India—IISc, IMT, MKU and University of Pune Web sites. The updates can be checked through 'What's new' and importantly comparative analyses of genomes, with advent of next generation sequencing (NGS), new genomes can be studied.

4. **Open access and ethics play a very important role for biologist. Do you agree? If so or not, how good are these to be exploited in Bioinformatics? Please give valuable suggestions.**

I slightly differ considering the fact that there should be closed access and some data may be kept confidential until it gets through the file of patenting records. Ethics also clubs to closed access. Thanks to the introduction of the new patenting regime in India!

5. **Making a successful principal investigator after years of post doctoral experience is a need by choice. Please provide suggestions with an example specific to development of bioinformatics in academia and industry.**

I would say that a PhD/Postdoc making a career front could go for tenure track positions which need not really affect his career. While, it need not be a choice, but it would essentially allow him to make a better PI. As discussed earlier it is not specific to bioinformatics but lest bioinformaticist work for both biology and information technology principles.

Pawan Dhar writes

1. How should bioinformaticist make up the blend of being a biologist as well as a programmer?

Bioinformatics offers an opportunity to help understand biology more accurately. What is accurate can be programmed. What can be programmed, can be understood. Thus, bioinformatics is a reasonable way to blend biological concepts and programming tools to help understand biology better. In my opinion, first students must identify widest variety of data that makes an organism. Second, students must know the biological context in which the data makes sense. Third, they must know the right mathematics and computational tools to play with the data. Finally, attempt should be made to build models that represent biology as we know it. The payoff is that if we understand biology, we can compose organisms. Students should clearly understand that if they are good biologists only then they can be good bioinformatician.

For bioinformatics students to become good biologists, they should be trained to write algorithms that accurately abstract biological processes, from molecular expression to cell–cell interaction and onwards. To appreciate the beauty of scale and complexity, it would be useful to train bioinformatician in experimental data collection—from sequencing to expression, structure, and flux measurement technologies and so on.

2. How best to formulate a problem for research in bioinformatics? What are the challenges one should pose?

To formulate a problem, following steps may be helpful.

Step 1: Read the discussion section of good scholarly papers published in the last 2–3 years.
Step 2: Make a document of unanswered questions. The questions usually appear in the form of direct statements or loosely thrown hints.
Step 3: Extract common and unique questions. The common ones are those that community are often discussing and may be important. The unique ones will be those that the authors are thinking along and could be important. One should expect to see some noise in such a data.
Step 4: Match the questions with your interests and available facilities, funding and so on.
Step 5: Add more questions with a hope to extend the intellectual front end of the field.
Step 6: Predict the value of answer (i.e., non-obviousness and scale), if one successfully addresses the problems.
Step 7: Assemble questions into an integrated research problem and imagine practical applications that would possibly emerge at the end of the project.

With reference to the challenges, there are at least two different approaches:

(1) For those who want to play safe, look for a challenge that would result in incremental but useful innovation.

(2) For those who like disruptive innovations, either (a) look for simple things that have been ignored, which, if properly addressed, may result in groundbreaking work, e.g., BLAST, or (b) scale up the work to an embarrassingly complex level, e.g., virtual organisms.

Since research is essentially an art to predict useful ignorance, students must be exposed to creativity and innovation exercises, in addition to getting trained in formal bioinformatics.

3. **Given the high-dimensional genomic data generated from time to time, how bioinformaticist should keep himself updated?**

Keep track of publications in key journals, follow conferences and symposia where speakers are likely to showcase latest unpublished results and make announcements. These days one need not physically attend an event as a number of talks appear online, both real-time and also as a repository.

4. **Open access and ethics play a very important role for biologist. Do you agree? If so or not, how good are these to be exploited in bioinformatics? Please give valuable suggestions.**

Yes, in my opinion for every scientist. Both open access and traditional models have advantages and limitations. However, given the financial crunch that publishers are facing in the traditional model, it seems to me that increasingly journals are going to opt for open access model in future. There are many bioinformatics journals that offer open access option. One may start more such journals provided focus is not just to make huge profits only but to identify the niche and maintain the quality, which in my opinion and experience is extremely difficult to balance. Given the fast paced competitive world that we are part of, I support the idea of 'first publish and then defend'. Another idea is to get the paper pre-reviewed by scientific community and publish it straightaway. I recognize that there are flaws in every publication model.

5. **Making a successful principal investigator after years of postdoctoral experience is a need by choice. Please provide suggestions with an example specific to development of bioinformatics in academia and industry.**

Following points apply both to academia and industry:

- Focus on algorithms more than tool development
- Move from sequence level to the pathway level and tissue level
- Find strategies to integrate every type of data that organisms offer
- Develop reliable literature mining tools such that one can build good molecular interaction models straight from the literature

- Maintain the database, if you have built one. However, maintain the database is infeasible, merge it with a more established larger database
- Design standards of data exchange and BioCAD tools for constructing organisms

Cox Murray writes

1. **How should bioinformaticist make up the blend of being a biologist as well as a programmer?**

Bioinformatics is really made up of three parts—biology, mathematics and statistics, and computer science. While bioinformaticists obviously need some coding ability, it is far more important that they have a good grasp of the biology and can frame biological questions as a logical series of analytical steps. Too many students with an interest in bioinformatics neglect vital areas of study, such as algorithms and statistics. A given problem can be coded in many ways, but a good understanding of algorithms can inform which are most efficient, or indeed, which are computationally tractable. Similarly, it is now easy to generate vast amounts of bioinformatics' data, but telling which patterns are meaningful remains much more difficult. It is increasingly important to have a solid grasp of statistics, including Bayesian and Monte Carlo approaches, especially as datasets get bigger and it becomes easier to misinterpret spurious, nonsignificant patterns in biological datasets.

2. **How best to formulate a problem for research in bioinformatics? What are the challenges one should pose?**

I would consider that there is no such thing as a bioinformatics' question. There are, however, biology questions, some of which can be addressed using computational approaches. The questions a bioinformaticist addresses should always be driven by the underlying biology. In this sense, it is important to distinguish whether a question can best be answered computationally or in the laboratory, or better yet, using a combination of both approaches. Some of the best bioinformatics' studies are those where computational and laboratory-based researches inform and support each other.

3. **Given the high dimensional genomic data generated from time to time, how bioinformaticist should keep himself updated?**

Good bioinformaticists require solid working skills in biology, mathematics and statistics, and computer science. Few researchers are equally conversant in all three areas. I advise my students to focus on up-skilling in their weakest subjects. It is also important to keep track of subject areas that are changing particularly fast, which in practice, unfortunately tends to be all three. It is often tempting to find quick, project-specific solutions to individual problems, but I encourage my students to discover generic solutions wherever possible. This means you already know the answer when you invariably encounter a similar problem again in the future.

4. **Open Access and Ethics play a very important role for biologist. Do you agree? If so or not, how good are these to be exploited in bioinformatics? Please give valuable suggestions.**

Issues around open access and ethics are not specific to bioinformatics. The best policy is to follow ethical guidelines for the associated biology field, which usually have the same broad underpinnings, but different specific requirements. Open access publishing is quite a different concern, and has its pros and cons. I am very supportive of making research results more accessible to the general public, as it seems unfair that taxpayers who have already funded research should have to pay again to access its results. Nevertheless, it is important to recognize that there are genuine, nontrivial costs associated with publishing, and these costs are not substantially lower for electronic-only publications. Open access publishing pushes these costs from the reader to the author, and in doing so, introduces new problems. Publishing costs, which can easily exceed US$2,000–3,000, limit who can afford to publish in Open access journals, and can often make researchers choose what research they want to release. Unfortunately, this divide is not obviously one between developing versus developed countries, but instead often falls within countries between well-resourced and poorly resourced research groups. Open access publishing is still an emerging phenomenon, and the progression and sustainability of this model are yet to fully play out.

5. **Making a successful principal investigator after years of post doctoral experience is a need by choice. Please provide suggestions with an example specific to development of bioinformatics in academia and industry.**

Although again not specific to bioinformatics, postdoctoral training is increasingly viewed as a necessary evil. The complexity of most modern research questions means that training beyond a PhD degree is required for most scientific jobs. I encourage students to expand their research horizons during their postdoctoral training by learning new skills unrelated to their PhD research. Training in a foreign country often proves extremely useful as well, particularly given the increasingly international nature of modern research programs. The PhD degree can now be considered basic training, while postdoctoral positions allow students to mature into well-rounded scientific researchers.

(i) Challenges and road ahead: Key skills and knowledge for bioinformatics
(ii) Bioinformatics as a career

Jeff W Bizzaro, President of bioinformatics.org opines

1. **How should bioinformaticist make up the blend of being a biologist as well as a programmer?**

Bioinformatics is cross-disciplinary. If your intention is to enter it from any one of the other STEM fields (Science, Technology, Engineering and Mathematics), you will need to supplement your education with courses or background material in certain places. For example, if your studies have or had a major focus outside of the life sciences, then you should also study the basics of biology, with an emphasis on genetics, genomics and proteomics. Likewise, if you have a strong background in the life sciences, you will need to learn about computer programming, databases, and statistical analysis.

However, just as it would be impossible for a biologist to become familiar with every single topic in biology, no one can expect to have a fully comprehensive education in bioinformatics. If you consider the fact that the field is at the intersection of several subjects, each requiring years of mastery, you will appreciate that the makeup of a bioinformatics research group is complementary by necessity. Once you do have a solid education in one major field and have completed some cross-disciplinary studies, it would therefore be best to quickly choose an area of interest within bioinformatics—a specialty.

2. **How best to formulate a problem for research in bioinformatics? What are the challenges one should pose?**

As a bioinformaticist, you are likely to collaborate with biologists and biochemists who do not have a background in programming. Nevertheless, any problem that involves programming can be described algorithmically, even employing some brainstorming sessions at a markerboard. At such times the common 'language' is science, and there may be little need to elaborate on the details of any programming that will be involved.

3. **Given the high-dimensional genomic data generated from time to time, how bioinformaticist should keep himself updated?**

You will find the greatest change in the available data if your interests include comparative genomics (the comparison of genomes between species) or newly investigated species. However, the genomic data for even the most highly studied organisms may already be somewhat static, in which case your research could involve lesser known aspects, such as gene regulation, epigenomics, protein–protein interactions, or protein structures.

4. **Open access and ethics play a very important role for biologist. Do you agree? If so or not, how good are these to be exploited in bioinformatics? Please give valuable suggestions.**

Ethics in bioinformatics can be thought of as a dichotomy between the need to reveal and the need to protect, perhaps more so than in any other field of science. On the one hand, science itself depends upon the free exchange of information, particularly for the purpose of reproducing and verifying results. Science as we know it cannot exist without such a collegial atmosphere.

One the other hand, individuals feels a need for privacy regarding their medical information, and so data that come from human trials or medical records need to be given special consideration. Along the same lines, companies and academic institutions also share an interest in profiting from discoveries and innovations and are often required by law to protect their information if they are to claim certain rights to them.

5. **Making a successful principal investigator after years of post doctoral experience is a need by choice. Please provide suggestions with an example specific to development of bioinformatics in academia and industry.**

A postdoctoral appointment is a necessary internship for those pursuing a career based on a cycle of grant proposals and subsequent funding. It could also be important in industry where the research and development phase of a product would be funded in a similar fashion. The role of a bioinformaticist, however, is oftentimes in support of larger projects that are not computational by nature. Drug discovery is an example of that. In such a case, a master's degree with several years of experience would be suitable to an employer.

Chapter 7
The Bioinforma 'TICKS': Frequently Asked Questions

1. What is bioinformatics?

Bioinformatics is a tool, whereas computational biology is a discipline. Bioinformatics predicts the biological outcome and can be used to compare the biological data, for example sequence analyses and structure prediction. In a nutshell, bioinformatics predictions can lessen the scale of experimentation. bioinformatics can be considered as a method to annotate the newly sequenced genomes. Bioinformatics can be well defined in biological and computational way.

Definition from biologists' perspective. Application of informatics and statistics to solve, analyze, annotate, and organize biological data in graphical and browsable formats.

Computational scientists' perspective. Design computational algorithms and applications for solving biological problems. By analyzing the existing biological data using information technology, we can predict the biological outcomes. Planning the analysis by workflow using bioinformatics tools and knowing expected output of the workflow will help to predict and solve the biological problems. Bioinformatics era has been started, and data are generated in huge amounts by next generation sequencing (NGS) in every field of biology, increasing the need of bioinformatics analysis.

2. Where can I be placed? Are there any companies working in the area of core bioinformatics research?

There are many institutes which require bioinformaticists to work with. After all, the role of bioinformaticist/bioinformatician is to help wet laboratory biologist plan his experiment or lessen his scale of experimentation using *in silico* methods. Say in India,

- Indian Institute of Science (IISc), Bangalore
- National Center for Biological Sciences (NCBS), Bangalore
- Institute of Bioinformatics and Applied Biotechnology (IBAB), Bangalore
- Jawaharlal Nehru University (JNU), New Delhi
- Jawaharlal Nehru Center for Advanced Scientific Research (JNCASR)
- National Institute of Mental Health and Neuro Sciences (NIMHANS).

P. N. Suravajhala, *Your Passport to a Career in Bioinformatics*,
DOI: 10.1007/978-81-322-1163-1_7, © Springer India 2013

And a host of all bioinformatics centers developed by Department of Bio-technology, Government of India.

3. Can I get placed in companies? Are there any companies that have bio-informatics resources/placements?

- Astrazeneca
- GVK Bio
- Aravind Bio
- Reddy's laboratories
- Ocimumbio.com
- Biocon

4. What are the branches/fields of bioinformatics?

(a) Computational Biology
(b) Drug Designing
(c) Phylogenetics
(d) Structural bioinformatics
(e) Population Genetics
(f) Genotype Analysis
(g) Systems Biology
(h) Synthetic Biology
(i) Functional Genomics

5. People say that bioinformatics has no scope. Is it true?

No, it is not so. Research is measured with publications, and now, almost all high impact factor journals are accepting with bioinformatics analysis in the articles. This shows the importance of bioinformatics in all fields of biology.

6. Do I need a programming experience to become a good bioinformaticist?

Yes, one do need programming experience for that. It helps to understand the most bioinformatics tools and their functionality; maybe you do not write your own programs, but surely it is an asset to learn more as a bioinformaticist.

7. How is chemo informatics different from bioinformatics?

Cheminformatics deals specifically with the chemical structures, whereas bioin-formatics deals with the biological systems and signaling pathways. Both the fields are devised so as to be able to manage huge data easily and come up with respective tools and techniques to study the same.

8. I am a B. Tech graduate. How can bioinformatics help me?

There are two options available: The first one is to go for M. Tech and then PhD, and the second option you can opt to work as project assistant with funded project or as a trainee/junior post with bioinformatics organization. If you like to have good grip on bioinformatics and start your career with good level, better go for first option.

9. Whither bioinformatics?

When it comes to bioinformatics, the biologist has the opinion of just storing the date or searching from different database such as doing the BLAST search, and now the things have changed with the time. As the different genome project moved up and algorithmic solution need with large data, thus biology itself has changed from a dogmatic, 'disciplinary' or 'pathway-based' science, to a broader, multi-disciplinary exercise.

10. What's new in bioinformatics?

According to Shankar Subramaniam of University of California, San Diego, there is a new 'central dogma'; genomes code for gene products, whose structures and functions are embedded in the pathways and physiology of biological activity. Each metabolic pathway can no longer be considered in isolation, but in the context of the interlocking and cross-coupled networks in which each component of that pathway participates. So the next solution is with a bioinformatician. According to Leroy Hood, founder of the Institute for Systems Biology in Seattle, such an approach not only needs a greater infrastructure (DNA/gene expression array technologies, proteomics, multiparameter cell sorting, mass spectroscopy, single cell assays, etc.) than traditional disciplines (molecular/cellular biology, biochemistry), but also requires advanced computational technologies. The major challenge the bioinformatics/system biology is facing for now is trained bioin-formatician and sufficient funding; here at Bioclues, we have taken the one challenge to have 2020 bioinformatician by 2020.

11. Who coined the term bioinformatics?

Paulien Hogeweg of University of Utrecht, Netherland, coined the word Theo-retical Biology in late 1980s.

12. How good the salary would be for a bioinformaticist?

It depends from one country to the other. As far as India is concerned, for a beginner, one can expect 15 k per month while medium-level scientists 40 k and a Senior level 60 k.

13. Can I do a PhD in bioinformatics? Where?

Of course you can. But please understand that bioinformatics is a tool. You may have to complement the wet-laboratory analysis done by someone or you need to liaise with a wet-laboratory biologist.

14. **Is there any integrated course/curriculum for bioinformatics?**

No. But IBAB, Bangalore has recently started the program. Many IISER institutes in India have recently started a 5-year integrated program catering to the needs of student scientists.

15. **Where can I undergo training after under graduation/graduation in bioinformatics?**

After completing your graduation/under graduation, try to seek a position in a reputed laboratory, which is working on bioinformatics. In this way, you would learn many tools and techniques which shall be adding to your profile.

16. **I want to do a project in bioinformatics. Can you suggest me?**

For doing a bioinformatics project, try approaching the people who are actually working on it in various research institutes by surfing the internet and writing those emails. Apart from this, you can enroll for live virtual project with Bioclues itself and get real-time problems to solve under the guidance of top notch scientists.

17. **If I take up M.Sc. bioinformatics, won't my area be more specialized and narrow? Do you suggest me to take up broader area for my Masters?**

Your subject would be more specialized as compared to any other broad field. No doubt about it, but you would be an expert in it. If you are confused as in whether bioinformatics is your cup of tea or not, then go for a broad subject for your masters in which you may study one paper on bioinformatics and later on pursue higher studies in the same to have the expertise.

18. **What is bioinformatics National Certification Program (BINC)? How does BINC help me?**

The BINC is certification/recognition exam, which will give the certificate for your skill in bioinformatics. This certificate will get a position with bioinformatics organization.

19. **I have done my B. Tech in bioinformatics. I am planning for my masters. I am confused between MS (Research) and M. Tech. I would like to know your valuable opinion on the career prospects of MS in Biotechnology and MS in bioinformatics as per the industry standards.**

We would suggest you to always go by 'interest' because opportunities reckon by interest not necessarily by choice. MS by research is a 'mentored' degree, and as a protege, you will be free to undertake a project of your interest. Typically, it lasts for 1.5 years with small amount of time dedicated to teach program. Both MS by research and M. Tech allow you to gain in-depth exposure to the component parts of bioinformatics. While the former focuses on research, the latter on pure taught program.

About the Author

Prashanth Suravajhala is a Post Doctoral Scientist and a virtual entrepreneur who founded Bioclues.org in 2005. He is also serving as an Associate Director of bioinformatics.org and has wide interests in lieu of *Functional Genomics and Systems Biology of Hypothetical Proteins in Human, specifically targeted to Mitochondria*. He loves mentoring undergrads who want to pursue bioinformatics. He can be reached at http://wiki.bioinformatics.org/prash

P. N. Suravajhala, *Your Passport to a Career in Bioinformatics*,
DOI: 10.1007/978-81-322-1163-1, © Springer India 2013

Bioinformatics Cross Word

Bioinformatics

Mind game

ACROSS	
4	Primer primer
5	Homologs duplicated
6	Well-known sequence format
8	NCBI
10	Orthologous sets of interacting proteins
DOWN	
1	Thermal cycler
2	A database but predicts
3	An alternative to Antibodies
7	Conserved sequences
9	Cluster of computers virtually

P. N. Suravajhala, *Your Passport to a Career in Bioinformatics*,
DOI: 10.1007/978-81-322-1163-1, © Springer India 2013

Epilogue

Through this book, the author overtly conveyed to the readers from day-to-day experiences, he has traversed whence his illustrious travel to several countries. Otherwise, the author hopes it tried to justify and confirm the traditional way of providing guidance to the beginners in Bioinformatics. Thank you for reading.

P. N. Suravajhala, *Your Passport to a Career in Bioinformatics*,
DOI: 10.1007/978-81-322-1163-1, © Springer India 2013

References

Altman, D.G., Bland, J.M.: Correlation, regression and repeated data. BMJ **309**, 102 (1994)

Bruggeman, F.J., Westerhoff, H.V.: The nature of systems biology. Trends Microbiol. **15**(1), 45–50 (2007)

Calvo, S., Jain, M., Xie, X., Sheth, S.A., et al.: Systematic identification of human mitochondrial disease genes through integrative genomics. Nat. Genet. **38**, 576–582 (2006)

Chen, Y., et al.: Variations in DNA elucidate molecular networks that cause disease. Nature **452**, 429–435 (2008)

Chotani, G., et al.: The commercial production of chemicals using pathway engineering. Biochim. Biophys. Acta. **1543**(2), 434–455 (2000)

Claudino, W.M., Quattrone, A., Biganzoli, L., Pestrin, M., et al.: Metabolomics: available results, current research projects in breast cancer, and future applications. J. Clin. Oncol. 1, **25**(19), 2840–2846 (2007)

Cornell, M., Paton, N.W., Oliver, S.G.: A critical and integrated view of the yeast interactome. Comp. Funct. Genomics. **5**(5), 382–402 (2004)

Cornish-Bowden, A., Cárdenas, M.L., Letelier, J.C., Soto-Andrade, J.: Beyond reductionism: metabolic circularity as a guiding vision for a real biology of systems. Proteomics **7**(6), 839–845 (2007)

Dell, H.G., Scott, R., et al.: A Greek-English Lexicon (1996)

Drummond, D.A., Bloom, J.D., Adami, C., Wilke, C.O., et al.: Why highly expressed proteins evolve slowly. Proc. Natl. Acad. Sci. USA **102**(40), 14338–14343 (2005)

Edwards, D., Stajich, J., Hansen, D. (eds.): Bioinformatics Tools and Applications, vol. XII, p. 451, 90 illus. Springer, New York (2009)

Fiehn, O.: Metabolomics—the link between genotypes and phenotypes. Plant Mol. Biol. **48**(1–2), 155–171 (2002)

Fiehn, O., Kristal, B., Van Ommen, B., Sumner L.W., et al.: Establishing reporting standards for metabolomic and metabonomic studies: a call for participation. Omics **10**(2), 158–163

Fierz, W.: Basic problems of serological laboratory diagnosis. Methods Mol. Med. **94**, 393–427 (2004)

Figeys, D.: Combining different 'omics' technologies to map and validate protein–protein interactions in humans. Brief. Funct. Genomic. Proteomic. **2**(4), 357–365 (2004)

Fox Keller, E., Harel, D.: Beyond the gene. PLoS ONE **2**(11), e1231 (2007)

Gomase, V.S., Tagore, S.: Physiomics. Curr. Drug Metab. **9**(3), 259–262 (2008)

Govorun, V.M., Archakov, A.I.: Proteomic technologies in modern biomedical science. Biochemistry (Mosc.) **67**(10), 1109–1123 (2002)

Greenbaum, D., Luscombe, N.M., Jansen, R., Qian, J., et al.: Interrelating different types of genomic data, from proteome to secretome: 'oming in on function'. Genome Res. **11**(9), 1463–1468 (2001)

Gronow, S., Brade, H.: Lipopolysaccharide biosynthesis: which steps do bacteria need to survive? J. Endotoxin Res. **7**(1), 3–23 (2001)

Han, X., Gross, R.W.: Global analyses of cellular lipidomes directly from crude extracts of biological samples by ESI mass spectrometry: a bridge to lipidomics. J. Lipid Res. **4**(6), 1071–1079 (2003)

Huang, S., Wikswo, J.: Dimensions of systems biology. Rev. Physiol. Biochem. Pharmacol. **157**, 81–104 (2006)

Huang, S.: Back to the biology in systems biology: what can we learn from biomolecular networks? Brief. Funct. Genomic. Proteomic. **2**(4), 279–297 (2004)

Hucka, M., Finney, A., Bornstein, B.J., Keating, S.M., et al.: Evolving a lingua franca and associated software infrastructure for computational systems biology: the systems biology markup language (SBML) project. Syst. Biol. (Stevenage) **1**(1), 41–53 (2004)

Kaeberlein, M.: Aging-related research in the "-omics" age. Sci. Aging Knowledge Environ. **4**(42), pe39 (2004)

Kasper, L., Karlberg, E.O., Størling, Z.M., Ólason, P.Í., et al.: A human phenome-interactome network of protein complexes implicated in genetic disorders. Nat. Biotechnol. **25**, 309–316 (2007)

Katam, R., Basha, S.M., Suravajhala, P., Pechan, T.: Analysis of peanut leaf proteome. J. Proteome Res. **9**(5), 2236–2254 (2010)

Khanna, C.: The dog as a cancer model. Nat. Biotechnol. **24**(9), 1065–1066 (2006)

Kim, T.Y., Sohn, S.B., Kim, H.U., Lee, S.Y.: Strategies for systems-level metabolic engineering. Biotechnol. J. **3**(5), 612–623 (2008)

Kuiper, H.A., Kleter, G.A., Noteborn, H.P., Kok, E.J.: Assessment of the food safety issues related to genetically modified foods. Plant J. **27**(6), 503–528 (2001)

McGuire, J.N., Overgaard, J., Pociot, F.: Mass spectrometry is only one piece of the puzzle in clinical proteomics. Brief. Funct. Genomic Proteomic **7**(1), 74–83 (2008)

Mehta, T.S., Zakharkin, S.O., Gadbury, G.L., Allison, D.B.: Epistemological issues in omics and high-dimensional biology: give the people what they want. Physiol. Genomics **28**(1), 24–32

Mons, B., Ashburner, M., Chichester, C., van Mulligen, E., et al.: Calling on a million minds for community annotation in WikiProteins. Genome Biol. **9**(5), R89 (2008)

Morel, N.M., Holland, J.M., van der Greef, J., Marple, E.W., et al.: Primer on medical genomics. Part XIV: introduction to systems biology—a new approach to understanding disease and treatment. Clin. Proc. **79**(5), 651–658 (2004)

Morrison, N., Cochrane, G., Faruque, N., Tatusova, T., et al.: Concept of sample in OMICS technology. Omics **10**(2), 127–137 (2006)

Oldiges, M., Lütz, S., Pflug, S., Schroer, K., et al.: Metabolomics: current state and evolving methodologies and tools. Appl. Microbiol. Biotechnol. **76**(3), 495–511 (2007)

O'Rourke, N.A., Meyer, T., Chandy, G.: Protein localization studies in the age of 'Omics'. Curr. Opin. Chem. Biol. **9**(1), 82–87 (2005)

Proceedings of the 3rd World Congress on Alternatives and Animal Use in the Life Sciences, Bologna, Italy, 29 Aug–2 Sept 1999

Reichert, A.S., Neupert, W.: Mitochondriomics or what makes us breathe. Trends Genet. **20**, 555–562 (2004)

Rochfort, S.: Metabolomics reviewed: a new "omics" platform technology for systems biology and implications for natural products research. J. Nat. Prod. **68**(12), 1813–1820 (2005)

Schloss, P.D., Handelsman, J.: Biotechnological prospects from metagenomics. Curr. Opin. Biotechnol. **14**(3), 303–310 (2003)

Schork, N.J.: Genetics of complex disease: approaches, problems, and solutions. Am. J. Respir. Crit. Care Med. **156**(4 Pt 2), S103–S109 (1997)

Steinfath, M., Repsilber, D., Scholz, M., Walther, D., et al.: Integrated data analysis for genome-wide research. EXS **97**, 309–329 (2007)

Stransky, B., Barrera, J., Ohno-Machado, L., De Souza, S.J.: Modeling cancer: integration of "omics" information in dynamic systems. J. Bioinform. Comput. Biol. **5**(4), 977–986 (2007)

Strömbäck, L., Jakoniene, V., Tan, H., Lambrix, P.: Representing, storing and accessing molecular interaction data: a review of models and tools. Brief. Bioinform. **7**(4), 331–338 (2006)

Suravajhala, P.: Hypo, hype and 'hyp' human proteins. Bioinformation **2**(1), 31–33 (2007)

Taylor, D.L., Woo, E.S., Giuliano, K.A.: Real-time molecular and cellular analysis: the new frontier of drug discovery. Curr. Opin. Biotechnol. **12**(1), 75–81 (2001)

Tracy, R.P.: 'Deep phenotyping': characterizing populations in the era of genomics and systems biology. Curr. Opin. Lipidol. **19**(2), 151–157 (2008)

Ward, N.: New directions and interactions in metagenomics research. FEMS Microbiol. Ecol. **55**(3), 331–338 (2006)

Werner, T.: Proteomics and regulomics: the yin and yang of functional genomics. Mass Spectrom. Rev. **23**(1), 25–33 (2004)

Web References

http://business.highbeam.com/138475/article-1G1-15682995/disciplining-evolutionary-biology-ernst-mayr-and-founding

http://www.nature.com/news/2011/110831/full/477020a.html

http://www.rncos.com/Report/IM382.htm

www.genecards.org